RAF HENDON
The Birthplace of Aerial Power

Front cover
Vickers Virginia bombers of 7 Squadron at the 1928 RAF Display.

Back cover (from top)
A Chipmunk, Anson and Dove in front of the control tower, the former Grahame-White company offices, in 1957.

A Spitfire of 1416 Flight in 1941.

The staff and equipment of the Hall School of Flying, circa 1915.

Reginald Carr stands in front of an Avro 504, 1919.

RAF HENDON

The Birthplace of Aerial Power

ANDREW RENWICK

RAF Hendon
The Birthplace of Aerial Power

Andrew Renwick
Crécy Publishing Ltd

Published in 2012 by Crécy Publishing Limited

A CIP record for this book is available from the British Library

ISBN 9 780955 426865

Printed in China

Crécy Publishing Limited
1a Ringway Trading Estate
Shadowmoss Road
Manchester M22 5LH

www.crecy.co.uk

CONTENTS

Introduction 6

1 **Claude Grahame-White** 8

2 **The London Aerodrome** 19

3 **The First World War** 38

4 **Flying Training** 41

5 **Aircraft Acceptance** 48

6 **Aircraft Production** 54

7 **Post-war Hendon** 61

8 **RAF Hendon** 74

9 **The Second World War** 82

10 **Peace Returns** 91

11 **The RAF Museum** 101

Postscript
The Grahame-Whites after Hendon....... 106

Acknowledgements 109

Index 110

INTRODUCTION

This book's subtitle was a phrase used by Claude Grahame-White and Harry Harper in 1919. While some would challenge the accuracy of the statement, Hendon has been at the forefront of many developments in aviation. Air mail, armament, aerobatics, aircraft manufacture, air safety and early air transport are just a few, together with education, for the pilot, the engineer and, most importantly, the public.

I have lost track of how many friends and acquaintances who served in Britain's armed forces said they had flown into RAF Hendon, or occupied various vantage points to watch the air displays. There isn't space in this book for a detailed history of the Hendon displays, the people associated with the aerodrome or RAF Hendon and its many units and personnel. Nor is it a comprehensive history of the RAF Museum, with its other sites at Stafford, Cosford and former restoration centre at Cardington. What I hope the book will do, however, is show the important role of the aerodrome at Hendon for more than 100 years.

Hendon's links with aviation, however, go back even further. On 20 August 1862 Henry Tracey Coxwell took a party of James Glaisher, Glaisher's son, Captain Percival and Mr Ingelow on a balloon flight. Glaisher often flew with Coxwell for meteorological research. On some flights, such as this, he was a paying passenger and took a reduced set of instruments. The flight suffered from light winds and they only just reached a field near Mill Hill where they landed for the night. Coxwell remained with the balloon while the others stayed overnight in the Greyhound Inn. They set off to walk the mile to the balloon at 4.00am and took off once more at 4.30, eventually reaching Biggleswade.

Most subsequent balloon flights were made from land near the Welsh Harp, with mixed results. Percival Spencer made a number of

Claude Grahame-White at the wheel of his Standard car with Harry Harper, aviation correspondent of the Daily Mail, *in the back. Together they wrote many books and promoted aviation.*

successful flights, including the first flight of the balloon 'Graphic' on 14 May 1902. Before that, however, others had less success. In September 1880 a race was organised by the Balloon Society of Great Britain. Mr Adams was one of ten competitors and was supposed to start from the Welsh Harp. Unfortunately the balloon only reached an altitude of 100 feet and landed about half a mile from its starting point. In August 1892 Corsican Louis Henri Capazza almost caused a riot. He was supposed to make a balloon ascent but the balloon slipped out of its net and took off without him. At this point the crowd turned angry and tried to attack him.

In the years following the first flight of the Wright brothers' aeroplane others began to experiment with them. In 1908 Helmuth Paul Martin and George Harris Handasyde began work on their monoplane at the Welsh Harp, close to Handasyde's home in Cricklewood. Attempts to fly were made near Edgware before they moved to Brooklands in 1910. By then, however, two electrical engineers had begun work on a new monoplane and an aerodrome from which it could be flown. Before they could use it, however, the aerodrome was used by the winner of the *Daily Mail* London-Manchester race. This event brought fame to the losing competitor, Claude Grahame-White, and laid the foundation for the development of the facility that subsequently became the London Aerodrome.

Below left: *James Glaisher was the first man to link Hendon and aviation, when he described a balloon flight that involved landing and taking off from a field in the parish.*

Below: *Helmut Paul Martin and George Harris Handasyde.*

Bottom: *Crowds gather at Hendon to watch Paulhan's Farman being assembled before his flight to Manchester, 27 April 1910.*

1 Claude Grahame-White

Before we consider the creation of the London Aerodrome we should consider the background of the man who made it famous – Claude Grahame-White. His family background is shrouded in uncertainty. Many claims are contradicted by other evidence and it was only in later years that the family adopted the name Grahame-White.

Ada Beatrice Grahame-White (née Chinnock).

On 11 May 1875 John White, Gentleman and son of John White, married Ada Beatrice Chinnock in Kensington, London. She was the daughter of Frederick Chinnock and Ellen Shackel, born in London on 26 April 1854 and baptised on 18 May at All Souls Church, Marylebone. Her elder sister, Florence, later married the Yorkshire businessman Francis Willey, about whom we will hear more shortly.

Beatrice Eleanor Genevieve was the eldest of the Whites' three children, born in Southampton in 1876. Little is known about her but she is often mentioned in connection with her brothers' activities. Claude's brother, Montague Reginald G., was born in Hound near Southampton in 1877. Claude was the youngest of the three children and was born on 21 August 1879. The date cannot be proven, however, because it has proved impossible to obtain a copy of his birth certificate.

By 1881 the family was living at The Towers, later known as Bursledon Towers, which was built by John White and is now the site of a supermarket. John made his money from property, enough to be able to employ four servants. Graham Wallace, Claude Grahame-White's biographer, states that he 'was a keen yachtsman, well known at Cowes, with a number of schooners in which the whole family cruised from April to September each year.' He also states that their mother employed French and German governesses so that they became fluent in those languages as well as their own.

Montague and Claude were both educated at a prep school in Crondall, Farnham, and at Bedford Grammar School. It had been planned that Montague would pursue a career in the Army but failed to be accepted for officer training and was forced to take a job in insurance. While working in London he saw his first motor car and became interested in this new form of transport, witnessing the 1896 Emancipation Run from London to Brighton. As the owner of one of the first cars he would have taken part if it had been working. He managed to secure employment with Daimler and in 1900 he accompanied Claude Johnson while surveying the proposed route of the Thousand Mile Reliability Trial. He also drove in the event, though not as a competitor.

Montague was a founder member of the Automobile Club of Great Britain. He made a name for himself by competing in the 1902 Gordon Bennett Cup for cars, which was run in conjunction with the Paris-Vienna Race. He shared a 30hp Wolseley with its designer, Herbert Austin. It was unfortunate that the rule requiring that all components had to be made in the country of the entrant led to mechanical failure. Montague subsequently featured in a film shown at the Palace Theatre, London, and it was here that he met the woman who would become his wife, Miss Florrie 'Birdie' Sutherland, a popular actress.

Not all of Montague's motoring exploits were received so well. On 10 May 1902 he took part in the Automobile Club's event at Dashwood Hill. Present was the Prime Minister, Arthur Balfour, and the club expected members to behave accordingly. Unfortunately Montague drove his 28hp Mors downhill at an estimated 60mph when the speed limit was 12mph. He was disqualified and suspended from future events and it appears he never raced again.

Claude gained an apprenticeship at a Bedford company in 1895, spending time in a bicycle factory building his own cycle. Both he and his brother competed in cycle races and, like his brother, his interests turned to motoring when he witnessed the 1896 London to Brighton run. Claude acquired a 3½hp Léon Bollée and became a founder member of the Automobile Club of Great Britain & Ireland. In 1898 he took his father in the car on the club's official run at Easter. Shortly after this he drove to Blyth Hall, where he converted his uncle, Francis Willey, to the benefit of motoring, before joining his company, the Shipley Wool Combing Co, where he introduced electric lighting to a new combing mill.

Far left: *John Reginald Grahame-White, the man responsible for changing the family name.*

Left: *Claude with his sister Beatrice and brother Montague. While the boys went to grammar school, Beatrice would have received most of her education at home.*

Montague on the Automobile Club Thousand Mile Trial of 1900; he is driving the Daimler support car in front of competitor 47.

Claude's next projects weren't an immediate success. He believed the lorry was more efficient than horses and persuaded his uncle to purchase a Lifu steam-powered lorry, which he collected from the works on the Isle of Wight and drove to Bradford. Unfortunately it skidded on the cobbled streets and shed its load on its first run. Despite this unfortunate setback lorries did indeed replace horses in due course. He left his uncle's company and became works manager for the Yorkshire Motor Vehicle Co at the Vaughan Motor Works, Bradford. He also intended to join the Reyrol Motor Car, which was formed in 1900 and planned to start car production at the Vaughan Motor Works, but failed to attract investors. At the end of 1901 he took a holiday in France over Christmas and the New Year.

Claude returned to England in 1902 and met George Wilder, who commissioned him to purchase a number of cars at the Paris Salon and deliver them to his estate at Stanstead Park, Sussex. As well as teaching him and his friends to drive, Claude was persuaded to become the estate manager while George and his wife, Una Evelyn Masie, an actress from Brooklyn, undertook an 18-month world tour. In 1905 Claude tired of managing Stanstead Park and left for an extended tour of South Africa, returning to Southampton on the *Armadale Castle* in April 1906. It is claimed that on the way out he was offered a job as an actor by the manager of a Gaiety Theatre company, which was to perform in South Africa; the offer was declined.

Back in England Claude's competitive side was tempted by new forms of motorised sport. Motor racing started on public roads in France in 1895, but was banned from roads in England, even after the repeal of the Red Flag Act. Only when Brooklands opened was motor racing in England possible, the events organised in a manner similar to horse races. He entered the second race meeting on 20 July 1907, but competed in only one race, the Hollick Selling Plate. He came second on a 27.9hp Minerva to A. Huntley Walker in a 34hp Darracq. It cost him 10 sovereigns to enter but his prize was 50 sovereigns. In August 1908 he entered *Carissima* in the British Motor Boat Club's races, held as part of the Cowes Regatta. On 3 August he won the fourth race, which was for boats that would not exceed 10 knots. On 7 August he came second in the race for a silver challenge cup for yachts, launches and boats of 10 knots and under.

In 1910 Claude wrote the draft of a speech while on board the White Star liner SS *Cymric*, sailing between Liverpool and Boston. Part of this speech is in the RAF Museum Archives as B768, and it is from this that the following quotes are taken. For 'several years general mechanical and electrical engineering attracted my attention and it was not until about 3 years ago [i.e. in 1907] that I renewed my interest in aeronautics. I then purchased a balloon of about 40,000 cubic feet and made several ascents from the Battersea gas works.' This purchase may have been made possible by inheriting a share of his late father's estate in 1906. In the same year the first Gordon Bennett Cup for balloons was held in France, which raised the profile of the sport.

In 1909 Claude Grahame-White registered C. Grahame-White & Co Ltd, automobile and marine motor engineers, with an office in 1 Albemarle Street, Piccadilly, and works at The Broadway, Walham Green (now Fulham Broadway). An idea of his character is suggested by notices he had in his office: 'Do it now' and 'Hustle like hell'. By now Claude had become bored with ballooning and sold his balloon. It is suggested by Montague, however, that their sister, Beatrice, maintained an interest in ballooning, having several flights with Henri de la Vaulx, eventually gaining her certificate of proficiency in France in 1911. Claude's interests lay firmly with powered craft, whether they were cars or boats; he kept the launch *Amoureuse* on the Thames at Maidenhead and had others on the South Coast.

Prompted by Wilbur Wright's 1908 demonstration flights in France and Blériot's cross-Channel flight in 1909, Claude went to the first aviation meeting at Rheims in August 1909. Despite being a member of both the Aero Club of Great Britain & Ireland and the French Aero Club, he was refused admission and had to watch events from a distance. This annoyed him but on the next day he succeeded in tricking his way in. He met Blériot, Farman and Levavasseur (sic) and agreed a contract with Blériot to buy his two-seat Type XII. *Flight* quoted him as also ordering an Antoinette monoplane, but nothing is known of it. He also placed contracts for the eight-cylinder ENV engine, and by February 1910 had acquired an exclusive agency in the UK for Chauvière propellers, examples of which he planned to exhibit at the Olympia show in March.

The Bleriot XII that Claude was to purchase was wrecked at Rheims by Blériot, so the construction of a replacement began. Blériot granted permission for him to go to the works and supervise the construction provided he placed himself under the discipline of the works manager and did not make any alterations. After three months the machine was ready and he received it on 6 November 1909. Next day Claude was impatient and began testing the machine by ground-running it, assisted by a friend; eventually they managed a short hop. Blériot was giving a demonstration in Vienna, but when he heard of their actions he informed them that the field was too small and that they should move the machine to Pau.

The group arrived at Pau about ten days later but the aircraft was damaged while being off-loaded at the station. It was repaired and Blériot arrived in time to fly it with Claude as passenger. During the flight it crashed and Claude was offered two single-seat Bleriot XI aircraft in exchange, both fitted with 25hp three-cylinder Anzani engines. This was agreed and delivery took place the next day. Blériot returned to Paris so Claude flew one of the monoplanes alone. He made rapid progress but managed to crash one in the afternoon. 'This somewhat unnerved me and I never yet have regained the same confidence I had prior to the accident.'

Ballooning had been popular for many years but for Claude it lacked excitement.

Grahame-White at Pau preparing his Bleriot XII.

Claude Grahame-White on a Farman biplane; the performance of European aircraft was superior to American aircraft at the time.

The extension of the tram service from Cricklewood to Edgware in 1904 improved public transport in the area.

Claude Grahame-White qualified for his French pilot's licence at Pau in December 1909 and it was awarded on 4 January 1910. According to the French Aero Club he is supposed to have passed the tests in a Farman biplane, but the Royal Aero Club recorded it as a Bleriot monoplane when it issued his British certificate in April 1910. After he received his licence he spent a week in London recruiting pupils for his school, which he opened at Pau.

In January 1910 Claude's flights in a Bleriot at Hendon were widely reported. Most accounts, however, come from a single source and are repeated verbatim in papers across the country. It has since been assumed that this was the site of what became the London Aerodrome, but that assumption may be wrong. On 20 January the *Evening News* reported that he made three flights of about half a mile at The Hyde, Hendon. Graham Wallace also says that Claude saw the site of his new aerodrome at a place called The Hyde, but this was a small settlement along the Edgware Road, just north of the Welsh Harp. An account published in *The Aero* for 18 January 1910 gives us vital clues and differs from all the others. In it the field is described as 'within half a mile or so of Hendon Station on the Midland Railway, that the tramline from Cricklewood runs within a hundred yards of the ground and that a new tube station is to be made two or three hundred yards from the entrance.' Even if one ignores the location of the planned tube station this implies it must have been near Edgware Road and closer to the Welsh Harp.

In February 1910, just after Claude had qualified for his French licence, *The Motor* reported the construction of six Bleriot monoplanes at Walham Green for the Grahame-White school. The works were on The Broadway, Walham Green, but were housed in little more than a large garage, making its exact location difficult to pinpoint. Work began following the appointment of Robert Wellesley Anthony Brewer, member of the Institute of Mechanical Engineers and author, as manager in January 1910. The aircraft were expected to be ready by March, once suitable engines were obtained. One of the Bleriots was sold to Bertie Fulton of Bulford Camp. *Flight* also reported on 5 February that Robert Brewer took a Bleriot from Hendon to Walham Green by road. It was reported that the aircraft had been flown successfully at Hendon prior to being dismantled.

While based in France Grahame-White's school had just one successful pupil. This was the American J. Armstrong Drexel, who, with William Edward McArdle, subsequently opened a school at Beaulieu, Hampshire. Another pupil was Edith Maud Cook (also known as Miss Spencer Kavanagh). She could have been the first British woman pilot if she had not been killed in an accident before passing her test; a parachute jump from a balloon in Coventry went wrong on 9 July and she died the next day.

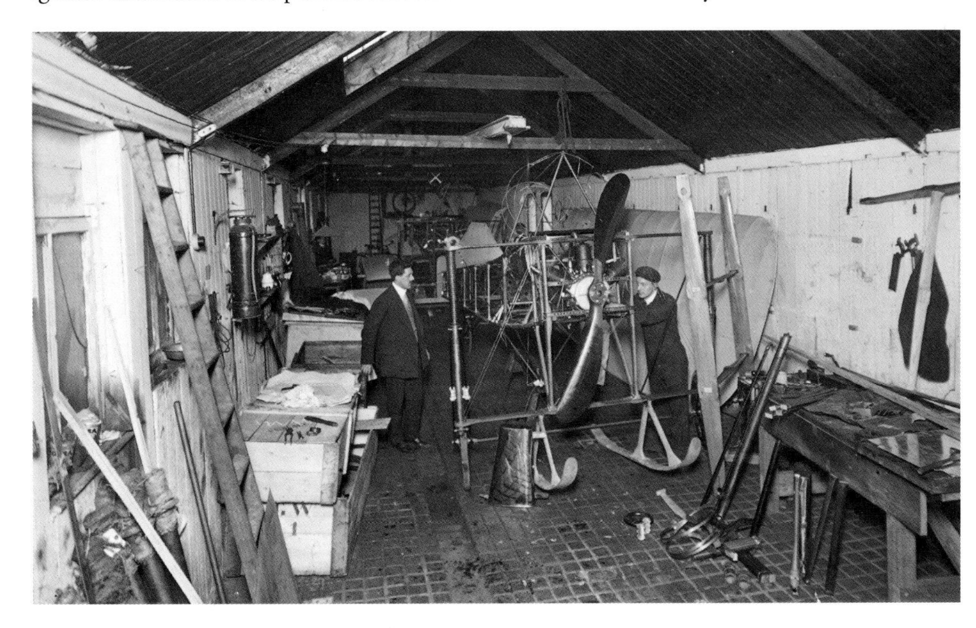

Six Bleriot XI aircraft were built at Walham Green under the supervision of Robert Brewer; they differed from others in having skids on the undercarriage.

Miss Spencer Kavanagh at Pau. She was the first British woman to learn to fly, but died before gaining her licence.

Lionel Henry Mander was the son of Theodore Mander of Wightwick Manor, Wolverhampton; the family is well known in the area and a shopping centre is named after them. He began learning to fly at Pau and moved with the Grahame-White School to Brooklands, where his flights in April 1910 were praised but he failed to qualify for his certificate. He subsequently became better known as the actor Miles Mander and was also a director, producer, playwright and novelist before his death in Los Angeles in 1946.

Claude became tired of flying around the airfield and began to fly across country, an unusual event at the time. On one occasion he was followed in a car by his mother, who had the misfortune to witness him crash. Another crash put him in hospital during March, but when he left later that month he travelled to Paris and bought a Farman. He took delivery at Chalons and flew it the same day, taking his mother for a flight on the following day. Almost immediately he arranged for the aircraft to be transported to London in order to compete for the London-Manchester prize, explaining his decision to the Farman brothers who tried to dissuade him. He found out from them that Louis Paulhan was also competing for the prize. They offered to split the prize money if successful in exchange for an improved machine for the attempt. Claude was surprised that he hadn't been sold the latest version, but eventually a deal was reached and he left for London to prepare for the flight.

Preparations for the London-Manchester flight had already been started by Robert Brewer. He had identified three possible locations for the start; Hendon, Park Royal and Wormwood Scrubbs. Brewer took a month's lease on Hendon, but it was rejected as being too open and it would be impossible to manage the expected crowd. Park Royal lacked a shed, but Wormwood Scrubbs had everything required. Claude went to Manchester to check for suitable landing grounds. On his return he found that his Farman was ready for shipping from Newhaven the next day. He had planned to use the *Daily Mail* 'garage', an airship shed, at Wormwood Scrubbs, but bureaucracy prevented him from obtaining permission. Eventually he took a short lease on the former Royal Agricultural Society showground at Park

Grahame-White's Farman biplane leaving Mourmelon on 12 April 1910 for the London-Manchester flight.

Royal and built a shed there. Just before he was due to make his flight he was granted permission to use a large recreation ground next to Wormwood Scrubbs.

Henry Farman arrived at Park Royal and checked the machine. Montague and Beatrice also arrived from Pau and agreed to follow the flight by car; Robert Brewer had gone on to the landing ground in Manchester. Bad weather prevented Claude flying from Park Royal to Wormwood Scrubbs, so he decided to start the next morning (Saturday 23 April) from Park Royal. The Museum's manuscript ends just after he refuelled at Rugby. High winds forced him to land at Hademore Crossing near Lichfield, where next day his aeroplane was blown over and damaged. Grahame-White took his damaged aeroplane to the airship shed at Wormwood Scrubbs and began the repairs needed. They were completed by the morning of Wednesday 27 April and Claude Grahame-White sought some sleep at the North Pole Tavern, Wormwood Scrubbs, before making a fresh attempt.

George Holt Thomas was born on 31 March 1870 at Hampton House, Clapham. He was the son of William Lason Thomas, who founded *The Graphic* newspaper later the same year. George became a Director of its sister paper, the *Daily Graphic*, and was heavily involved with the promotion of aviation in Britain. The paper offered the Graphic Prize in 1906 for the first flight of 1 mile, and George Holt Thomas was one of Louis Paulhan's sponsors. Paulhan had made a number of demonstration flights in England and had no worries about using Hendon for his attempt to win the London-Manchester prize. When Farman left Grahame-White on 27 April he went to Hendon where Paulhan's aeroplane had been delivered and awaited erection. Grahame-White was woken in the evening with the news that Paulhan had departed for Manchester, having taken off at 5.21pm.

Henry Farman and Claude Grahame-White checking the biplane for the London-Manchester flight.

Grahame-White had expected Farman to warn him when Paulhan was ready, but the note

Grahame-White's Farman on 24 April 1910 after it had been blown over at Lichfield.

Paulhan and his backers in Manchester. Left to right, Madame de Kersauson de Pennendreff, Henry Farman, Madame Paulhan, Louis Paulhan, Baron Robert de Kersauson de Pennendreff, and George Holt Thomas.

didn't reach Claude's supporters at Wormwood Scrubbs until after he too had taken off. Darkness forced Paulhan to land at Lichfield, and Grahame-White at Roade. Despite the darkness, Grahame-White took off once more, aided by the lights from the support cars, the first flight at night in England. Engine trouble, however, forced him down once more at Polesworth. Before Grahame-White could cure the problem, Paulhan had taken off once more and completed the flight to Manchester on Thursday 28 April, a flying time of 4 hours 12 minutes for the 195 miles and an average speed of just over 46mph.

In April 1910 Claude Grahame-White's name was mentioned in connection with another company, Aviation Investment & Research Ltd. While the directors aren't well known in aviation, some of those mentioned in connection with various committees are: Sir Hiram Maxim, J. T. C. Moore-Brabazon, T. W. K. Clarke, Warwick Wright and others, in addition to Grahame-White. The company was short-lived, however, and the decision to wind it up was taken on 15 November 1911, though this wasn't completed until 16 March 1916.

In May 1910 the aviation press reported that it was planned that Grahame-White's school at Pau would move to Park Royal. He was renting it as an aerodrome and plans had been passed for workshops and hangars. By then one of the six Bleriots his company had been constructing was tested at Park Royal. Pleasure flights and exhibitions were also planned. In fact, the school moved to Hangar 16 at Brooklands in April and stayed there; it was from here that Grahame-White did most of his flying at that time. In May he was prosecuted for speeding, and used his Farman to fly from Brooklands to the court in Woking.

The school's activities were curtailed by Claude's need to exploit his fame, earning money by providing flying demonstrations and entering competitions. His first demonstrations were at the Ranelagh Club, Barnes, but he was often grounded by the weather. He then moved to the Crystal Palace, where the Aerial League had built the Aerial Rendezvous, a cottage with a small aeroplane shed attached. Even though the lawn in front was only flat for 50 yards before a steep slope, Grahame-White was able to get airborne in just 22 yards on 7 June. After this he moved north to Halifax, from where he also visited the Northern Aero Syndicate at Bradford. He also competed in the Wolverhampton, Bournemouth and Blackpool aviation meetings.

On 6 July Claude Grahame-White attended a luncheon at the Savoy Hotel, during which

Grahame-White about to give a display at the Ranelagh Club, Barnes, in June 1910. The aircraft used was the Farman from the London-Manchester race.

Lady Abdy offered £50,000 to help establish an air link between London and Paris, and Claude was to be one of those to help organise it. Also present was Pauline Chase, the American actress famous for her role as Peter Pan, and god-daughter of James M. Barrie, creator of the character. Claude had first met her when she appeared in *The Girl Up There* at the Duke of York's Theatre in 1901, her first acting role in London.

On 9 October Claude Grahame-White and Pauline Chase were reported to have had lunch at Delmonico's in Boston. The report in the *New York Times* also indicated that she would be present at Belmont Park when her acting schedule permitted. Finally on 18 October Pauline Chase announced her engagement to Claude. She was to remain in New York until December, when she would return to the UK with him and continue to play Peter Pan until her marriage and retirement from the stage.

Grahame-White had sailed to the United States in August 1910 to compete in the Harvard-Boston and Belmont Park meetings. It was after the latter that his former pupil, Drexel, resigned from the Aero Club of America in protest at the treatment afforded Grahame-White and de Lesseps. One of his future pupils, however, was James Vernon Martin, a student at Harvard who helped form the Harvard Aeronautical Society, which organised the Boston meeting.

Grahame-White attracted a great deal of attention in the USA, not least when he landed in Washington DC on 14 October 1910. It was claimed that he met President William Howard Taft at the White House, but this was not so. He actually visited General James Allen of the Signal Corps, Major-General Leonard Wood

The actress Pauline Chase; her real name was Ellen Pauline Matthew Bliss.

and Admiral Dewey. The visit was organised by his friend Clifford B. Harmon as a response to visits by General Allen and Commodore John Barry Ryan when Claude Grahame-White was giving demonstrations at Benning race-track near Washington. Grahame-White had, however, met President Taft at the Boston meeting, where his son, Charles Taft, acted as Claude's assistant.

One person who left Grahame-White's company at this time was Monsieur André Grapperon. He had held many world records for motor cycles before becoming a pupil at Blériot's school in Pau. He joined Grahame-White as the 'Directeur' of his school at Pau and was Grahame-White's chief mechanic during the London-Manchester flight of 1910. In October 1910 he returned to Paris to start work on his own aeroplane before returning to motor cycle racing.

The Wright brothers were very protective of their developments, and in 1908 sued Glenn Curtiss, claiming that he had infringed their patent. The Wright Company was formed by a group of American businessmen and bought the Wright Patents for $100,000. They sued Claude and others because they imported aeroplanes that had not been licensed by them, despite the Aero Club of America having an agreement with the brothers. The writ was served on Grahame-White just before he returned to England. A similar writ was served on Louis Paulhan, but he joined Glenn Curtiss and challenged the validity of the Wright patents. Claude took no action, so the court assumed he accepted the validity of the Wright claim.

Grahame-White arrived back at Liverpool on 6 December 1910 on the SS *Mauritania* and was given a hero's welcome, having won the Gordon Bennett Cup for aeroplanes for Britain. Within a week of his return he travelled to Dover in search of another trophy, the Baron de Forest prize. He had ordered a Bristol Boxkite, which was delivered to him at Dover after being test flown by Maurice Tetard, the company pilot. After a couple of setbacks, including a dog being killed by the turning propeller and damage during a winter gale, the biplane was repaired. On 18 December he decided to try his attempt at winning the prize and took off from Swingate Downs despite strong winds. While turning, the aircraft was hit by a gust, side-slipped into the ground and was wrecked; although Grahame-White needed medical treatment, he planned another attempt before the close of the competition on 31 December 1910. On Christmas morning, however, a fire destroyed his replacement aircraft and brought the attempt to an end. Despite this he finished the year as the sixth most successful pilot in terms of winnings, and the highest-earning Briton; he earned $22,000 at Boston alone.

Grahame-White landing on Executive Avenue, Washington DC, 14 October 1910.

The London Aerodrome 2

Mr Edgar Isaac Everett, the son of a miller's foreman, was born in Metfield, Suffolk, in 1869. In 1892 he married Mary Jane Andrews in Hemel Hempstead and they raised four children. With Charles William Brown he formed Everett & Brown of 11 Albemarle Street until the partnership was dissolved on 13 May 1896. He subsequently formed Everett & Co of 22 Charterhouse Square with William Joseph Cole, until this partnership too was dissolved. His next partner, Kenelm William Edward Edgcumbe, was born in Vienna on 9 October 1873, the son of Richard Edgcumbe, Sergeant-At-Arms to Queen Victoria. He joined the Royal Engineers (Volunteers) and on 1 April 1908 was appointed Captain in the London Division (Electrical Engineers) Royal Engineers (Volunteers). It is claimed that he joined Everett in 1898, and together they formed Everett Edgcumbe & Co Ltd and premises were acquired in Colindeep Lane, Hendon. Their interest turned to aviation and they began work on a monoplane. Some say they were assisted by another Hendon-born engineer, Charles Richard Fairey, who it is also claimed worked with Martin and Handasyde at the Welsh Harp.

In order to test their monoplane an airfield was required. *The Motor* published a photograph on 1 February that showed trees being felled to clear a site for flying. In February 1910 Hendon Urban District Council approved the erection of a shed in which to house the Everett-Edgcumbe monoplane, and work could begin. In February 1910 it was also reported that Claude Grahame-White had successfully flown a Bleriot monoplane at Hendon. In March 1910 the pioneer aviator Robert MacFie made the point, however, that aircraft such as the Bleriot XI rose from the ground quickly and that 'probably half a mile square would do'. The field that Grahame-White viewed with Everett and Edgcumbe was supposed to be ready by April, but may have provided enough space for the Bleriot before the work was completed. The Everett-Edgcumbe design was not so efficient and would have to wait for the clearance of a larger area. Robert MacFie also wrote that Park Royal had 'not enough smooth ground to learn on' and that Hendon was 'very promising; not yet ready'. Despite this, Grahame-White chose Park Royal, and Louis Paulhan used Hendon for the London-Manchester race.

The Everett-Edgcumbe monoplane in its shed at Hendon in 1910; it might have flown with an engine of greater horsepower.

The size of an aerodrome was important. Airfields today accommodate runways long enough for aircraft to take off and land safely. In the early days of aviation, however, all flying was carried out over the airfield; pilots rarely ventured beyond the aerodrome boundary, which made cross-country flights remarkable. The area chosen for the aerodrome was defined by the railway line in the east, what is now Aerodrome Road in the south and a tributary of the Silkstream in the west; the northern boundary was less well defined and there was the possibility of expanding in this direction as far as the railway from Mill Hill East to Edgware.

On 30 September 1910 Kenelm Edgcumbe and his brother-in-law, Harold Arthur Arkwright, registered The London Aerodrome Company Limited to acquire the site. Leases were obtained and more ground was cleared; on 2 November 1910 Lawrence Ardern was made a shareholder in recognition of the assistance he provided in obtaining the various leases. Sheds were erected by F. Smith & Co of Stratford, East London, and although the flying ground wasn't completely ready it opened on 1 October 1910. *The Aero* reported that enough ground had been cleared to allow a run of half a mile, with workmen clearing more hedges and trees to create an area of more than 200 acres and allow a circuit of 1¾ miles. The secretary of the London Aerodrome, Mr Clutton, was reported as seeking more occupants, and that they had no intention of promoting 'gate money' meetings that would interfere with their work.

The sheds at Hendon initially housed just two organisations. The Bleriot School opened with Norbert Chereau as manager, Pierre Prier as chief instructor and Frank Hedges Butler as its first pupil. Chereau had come to Britain in 1893 and had been Blériot's representative in this country for many years, marketing Blériot's

Pierre Prier, chief pilot of the Bleriot School and the first man to fly at the London Aerodrome after it opened.

The Aeronautical Syndicate moved its operations from Larkhill to Hendon in September 1910, just before the aerodrome opened.

lamps for cars and assisting with the arrangements for the successful crossing of the Channel in 1909. The other organisation, the Aeronautical Syndicate Ltd (ASL), was registered by Horatio Barber in July 1909 for the design and construction of aircraft. It was based initially at Larkhill, but lacked space, a problem solved by moving to Hendon. Late in October the two companies were joined by the Howard Wright Monoplane of Captain Geoffrey Llewellyn Hinds-Howell ASC.

The aeroplanes of the Bleriot School and the Aeronautical Syndicate were regularly seen in the sky over Hendon at a time when few people had even seen a motor car. This was in contrast to the Everett-Edgcumbe monoplane. In November 1910 it was reported that the monoplane was taken out of its shed after a long period of reconstruction but broke a wheel and needed repairing once more. The aircraft never did succeed in flying, although a few hops were made by Everett and E. C. Clutterbuck. No more was written about the monoplane after January 1911 and its fate is unknown. Clutterbuck subsequently joined the ASL School.

As the Everett-Edgcumbe monoplane disappeared from view the Pupin monoplane made its appearance, arriving at Hendon in November 1910. It was still present in February 1911, but no mention is made of it ever flying. Emile Pupin went on to become a director of the Aerial Propulsion Syndicate Ltd in 1913.

Once more hangars had been built, the Grahame-White School was able to move from Brooklands and new workshops replaced those at Walham Green. Grahame-White recruited Clement H. Greswell, his former pupil, as instructor for the school; he had qualified on 15 November 1910 while Grahame-White was still in America. While in the States Grahame-White had also acquired an agency for the Burgess Model E Baby, which he imported, assembled and marketed as the Grahame-White Baby. At least seven aircraft were ordered and one was used at his school; another was fitted with a simple nacelle for the pilot, with the name 'Baby Grahame-White' on the nose. Grahame-White also acquired a licence to build Morane monoplanes in his new works at Hendon. In January 1911 the company was joined by Gordon Stewart, designer of the Mulliner monoplane, who took charge 'of the constructive branch at Hendon'. The company name and office, however, was still C. Grahame-White & Co Ltd of 1 Albemarle Street, Piccadilly. In January 1911 adverts began to appear for the Grahame-White School at Hendon and, once it was open, reports of its activities began to appear in the aviation press, firmly linking the two.

Claude had first visited Blackpool and Southport in August 1910. He returned to Freshfield, Southport, in February 1911, where he took the opportunity to meet with his fiancée, who was starring as Peter Pan at the Southport Opera House. Claude had been expected to marry her in the spring of 1911, but for some reason it did not take place. She continued acting until she married Captain

Workers inside the Grahame-White factory at Hendon.

Alexander V. Drummond of the West Kent Yeomanry on 24 October 1914.

On 12 April 1911 Pierre Prier made a non-stop flight from Hendon to Paris. This was remarkable, coming less than two years after Blériot's famous cross-Channel flight and just under twelve months since the exploits of Grahame-White and Paulhan in the London-Manchester race, and highlighted the improvements in performance and reliability of aircraft. Unfortunately Prier joined the British & Colonial Aeroplane Company as a pilot and designer in June 1911, so in October Henri Salmet was appointed as his successor; he had been taught to fly by Pierre Prier at the Bleriot School.

Hendon's first significant event was held in May 1911. The Chairman and Secretary of the Parliamentary Aerial Defence Committee, Arthur Lee and Arthur Du Cros, organised a

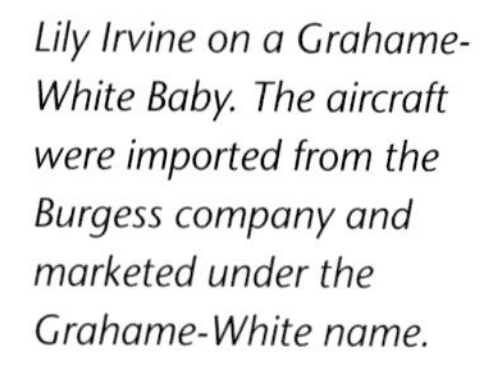

Lily Irvine on a Grahame-White Baby. The aircraft were imported from the Burgess company and marketed under the Grahame-White name.

Claude Grahame-White about to set off to provide a demonstration of bomb-dropping during the visit of the Parliamentary Aerial Defence Committee.

demonstration of aviation at Hendon on 12 May. They were assisted by Grahame-White and Blériot, but the Aeronautical Syndicate was prevented from participating, despite having suitable aircraft and having been resident at Hendon since it opened. The aviation press was disgusted by the exclusion, and it probably contributed to Horatio Barber's decision to leave. Despite his treatment, Barber presented four aircraft to the British Government, two each for the War Office and the Admiralty, before concentrating more on research. Management of the Aeronautical Syndicate passed to W. Ridley-Prentice, a student pilot at the Grahame-White School and member of the Royal Aero Club. Horatio Barber subsequently became a consultant, assisted by C. W. Harris, his former works manager.

In the event, the Government did not take delivery of any of the Aeronautical Syndicate

Howard Pixton flew over from Brooklands in the Avro Type D to participate in the Parliamentary display.

Valkyries, which consisted of two single-seat monoplanes with 30hp and 40-50hp Green engines, and two two-seat monoplanes with 60hp Green and 50hp Gnome engines. Two were to go to the Air Battalion and two to the Royal Navy. On 17 September 1911 Lieutenant Reginald Archibald Cammell was killed when the two-seat Valkyrie with a 50hp Gnome he was testing crashed at Hendon. Despite this setback the company went on to build the Viking biplane in 1912.

Although Claude Grahame-White had won the 1910 Gordon Bennett Air Race, the 1911 race was held on 1 July at the Royal Aero Club flying ground at Eastchurch. As early as January 1911 the *Morning Post* was one of those suggesting Hendon as the best possible location in Coronation year. Despite this snub, two races did include Hendon. In July both the European Aviation Circuit and *Daily Mail* Circuit of Britain races included Hendon as a control point. People had been generous to Claude Grahame-White when he lost the London-Manchester race, and in return he organised a benefit meeting for Jules Vedrines. Vedrines had come second to 'André Beaumont' (Lieutenant de Vaisseau Jean Conneau) in both races. An estimated 50,000 attended and watched Beaumont, Vedrines and Grahame-White perform; Vedrines left the event £2,000 richer and Hendon had witnessed its first public air display.

Following his earlier visits to Southport, Claude Grahame-White was persuaded to give a series of demonstrations as part of the Coronation Gala in June 1911. They were not a complete success but Leonard Williamson helped arrange a further series of demonstration flights. These ended in August with even less success than before because of bad weather. In return Williamson supervised the construction and fitting-out of nine sheds for Grahame-White, with completion at the end of October 1911. The first three were used for the construction of components that were assembled into complete aircraft in the fourth and fifth sheds; the other four were used for the school aircraft.

One of the sheds used by Grahame-White had previously been occupied by another tenant. In late February or early March 1911 the Chanter School was established at Hendon with two Bleriot XI aircraft in one of the sheds Grahame-White wanted for his school and factory. Not only was the owner of the school the instructor, but he was also a pupil himself. Hamilton Ross, who later became company manager, was also one of the pupils. There was a disagreement between the Chanter School and its new landlord, which led to it being banned from flying on 21 September 1911 and all pupils excluded from the aerodrome. A settlement was reached, however, and the school resumed training on 7 October, but by then Chanter had decided to relocate to Shoreham, which he did in November.

The argument with the Chanter School took place while Grahame-White was out of the country again. During 1911 he had moved from Westminster to 1 Red Ruth Villas in Colindale Avenue, giving him a home closer to the aerodrome. However, he cannot have spent much time there because of his other commitments, and had to depend on his staff to run his business. As soon as he had completed the acquisition of London Aerodrome and finished his flights in Southport, he returned to the United States with his sister on the *Mauritania*. In November he had no sooner returned to the UK than he was off to France, before returning to the USA once more on the SS *Olympic*, arriving on 6 December 1911.

While in the USA Claude took part in the Boston Meeting; he also planned to undertake exhibition flights and meet with James Martin. It was not long, however, before his failure to challenge the writ of 1910 came back to haunt him because a court granted the Wright Company an injunction, preventing him from flying in the USA until the legal case was resolved. A second action for damages was rejected by the court but the whole episode left Grahame-White depressed. He told the US media that he was considering giving up flying completely and concentrating on aircraft construction. His feelings were made stronger by the type of pilots engaged in exhibition flying and their high death rate; one of the worst offenders was the Wright Company itself.

Grahame-White was still out of the country when Hendon's final event of 1911 took place in September. Captain Walter George Windham had organised the world's first air-mail service from Allahabad to Naini Junction during the United Provinces Exhibition in

India in February 1911. He was persuaded to organise a second air-mail service from Hendon to Windsor to celebrate the coronation of King George V, an enthusiastic philatelist. The main coronation celebration was the Festival of Empire at the Crystal Palace; Grahame-White shareholder Herbert William Matthews was its business manager. Aircraft and pilots for the service were provided by the Grahame-White Aviation Co and the first flight was made by Gustav Hamel on 9 September. Claude Grahame-White estimated that 130,000 items had been carried by the end of the last flight on 26 September.

Little has survived to provide information about the acquisition of Hendon by Grahame-White. There is, however, a letter from Arkwright to Claude Grahame-White dated 13 March 1911. In it he acknowledges receipt of a letter of the same date exercising Grahame-White's option to buy the London Aerodrome and the deposit of £4,000 at the bank, possibly the result of an agreement dated 24 December 1910. This may have been a clause in the agreement between the London Aerodrome Ltd and Grahame-White when he leased the sheds at Hendon and moved his school there. On 12 June 1911 Hendon Council approved the use of a temporary building at the London Aerodrome as offices for C. Grahame-White & Co Ltd. This was the conversion of the Everett-Edgcumbe shed, which became his company's office from 24 June.

On 24 March 1911 Grahame-White and Blériot entered into an agreement with a new company that would merge their schools. In addition, Richard Thomas Gates signed another agreement with Claude Grahame-White on 25 March on behalf of the new company for the acquisition of the London Aerodrome. The new company was Grahame-White, Bleriot & Maxim Ltd, incorporated on 28 March 1911 with its office initially at 1 Albemarle Street. On 1 April 1911 a prospectus was issued for the sale of shares, but it was under-subscribed and Louis Blériot decided to have no further part in the enterprise; Sir Hiram Maxim also withdrew from the project. The first shareholders, therefore, were Grahame-White as Chairman, Richard Gates as General Manager, Lt William George A. Ramsay-Fairfax RN and Herbert William Matthews, an architect. Richard Gates and William Ramsay-Fairfax had both served in South Africa in 1902 as Lieutenants in 30 Battalion, the Imperial Yeomanry. At an extraordinary general meeting on 19 July the decision was taken to rename the company the Grahame-White Aviation Co Ltd, a decision confirmed on 9 August, and it was this

Evelyn Frederick Driver delivering the mail to Windsor.

company that soon became the proprietor of the London Aerodrome.

On 31 July 1911 there was an extraordinary general meeting of the London Aerodrome Ltd at which the decision was taken to sell the London Aerodrome to Claude Grahame-White. The decision was confirmed on 18 August and a liquidator was appointed to voluntarily wind up the company, the final meeting of which was held on 1 October 1912. On 3 August 1911 new leases were signed between Claude Grahame-White, William George Astell Ramsay-Fairfax and the land-owners for ten years from 29 September 1910. The lands were 146 acres of Church Farm from Colonel Sir Theodore Brinckman, 36 acres of Featherstone Farm from John Heal and Maurice & Cecil Brewer, and 24 acres of Tithe Farm from Sir George Barham.

The agreement of 25 March 1911 had lapsed, so a new one was signed on 11 August between Claude Grahame-White, C. Grahame-White & Co Ltd and George H. Mansfield, transferring the aerodrome to Grahame-White Aviation Co Ltd. At a meeting of 14 October 1911, confirmed on 30 October, the decision was taken for C. Grahame-White & Co Ltd to enter voluntary liquidation, with George H. Mansfield appointed liquidator; the task was finally completed in July 1915. One of the first things the new company did was to reduce the rents for the hangars in a bid to attract more tenants.

The sale of the aerodrome seems to mark the end of Everett & Edgcumbe's involvement in aviation. Nothing is known of Everett until his death in Sussex in 1934, but Edgcumbe served in the Army during the First World War, reaching the rank of Major by 1919. He became President of the Institute of Electrical Engineers and 6th Earl of Mount Edgcumbe before his death in 1965.

One of those Grahame-White had met at Southport was Cecil Compton Paterson, a director of the Liverpool Motor House Ltd. He helped establish the flying ground at Freshfield, Southport, and had his company build his biplane, which flew successfully on 14 May 1910. He met Claude at Freshfield in August 1910, and in December gained his licence before moving to London. He spent a lot of time flying for the Grahame-White School, but in October 1911 it was reported that he was using a field near the Welsh Harp for test flights of his latest biplane; it is unclear why he didn't use Hendon.

In December 1911 Cecil Compton Paterson moved to South Africa and established the African Aviation Syndicate Ltd at Cape Town. With him were South-African-born Evelyn Frederick Driver and Captain Guy Livingston, Grahame-White's former general manager (later Brigadier-General CMG RAF). Together they helped promote aviation in Africa, but it meant that Grahame-White had to recruit more staff to replace them.

Grahame-White's stand at the 1911 International Aero Show at Olympia, advertising the London Aerodrome and his new company.

The original aeroplane sheds at Hendon in June 1911; in the background the former Everett & Edgcumbe shed is being converted into offices.

Among the new recruits to the Grahame-White Aviation Co was John Dudley North, who had previously been employed by the Aeronautical Syndicate until its closure. The company now began to produce its own aircraft instead of assembling imports or repairing school aircraft. Grahame-White could also offer his services to the War Office as an aircraft constructor. His premises were inspected in 1912 and an order placed for two RAF BE.2a aircraft. When he sought more work in 1913 he was informed that no more orders would be placed until the first one was completed. In reply, a list of problems was presented, not least the late delivery of engines for the aircraft, a story that would repeat itself. Despite this, the War Office purchased seven existing aircraft from the company in March 1913, the so-called Grahame-White transaction.

The Aeronautical Syndicate had struggled after the departure of Horatio Barber, and it decided to close. It was planned to sell the company's assets at auction on 24 April 1912, but instead they were acquired by Fred Handley Page. He sold most of them on to George Holt Thomas, including the aeroplane sheds and workforce. Once the assets were gone the Chairman, Hermann Rudolf Schmettau, called an extraordinary general meeting on 30 December 1912 and the decision was taken to voluntarily wind up the company. This was confirmed on 16 January 1913 and a liquidator appointed. In 1911 George Holt Thomas founded The Aircraft Company, and by 1912 had acquired the exclusive licence to build and sell Farman aircraft and Gnome engines in

The Farman "Wake Up England" biplane. This aircraft was used in the 1912 exhibitions sponsored by the Daily Mail, *and the quote was from King George V when he was Prince of Wales.*

Lieutenant John C. Porte was manager of Deperdussin, one of the companies attracted to Hendon.

Sydney Pickles and a Caudron biplane.

A Breguet biplane in flight; in the background are the sheds that housed the representatives of Farman, Breguet, Caudron and Deperdussin.

England. The assets acquired from the failure of the Aeronautical Syndicate meant that the company could manufacture aircraft instead of just importing them, and on 6 June 1912 the company was renamed the Aircraft Manufacturing Co Ltd (Airco).

Airco had been fortunate by its acquisition of the Aeronautical Syndicate's sheds. The number of aeroplane sheds at Hendon was always a limiting factor, restricting the schools and manufacturers able to use the aerodrome. More were erected in order to accommodate the increasing number of companies, especially aircraft manufacturers, interested in a move to the area, and throughout 1912 new companies arrived, large and small.

The Blackburn School of Harold Blackburn was one of the smallest and moved to Hendon in September 1912. The school closed in mid-1913 when Blackburn became personal pilot to Dr Malcolm Grahame Christie, a former pupil of the school. Another small school was the Temple School, founded by George Lee Temple and Mr Jameson. As with M. Chanter and his school, Temple was a pupil of his own school; he taught himself to fly and gained his licence on 18 February 1913.

The Deperdussin School was originally formed by the British Deperdussin Aeroplane Syndicate Ltd at Brooklands in 1911, with John Cyril Porte as managing director and chief instructor. In 1912 that company was liquidated, its assets acquired by the British Deperdussin Aeroplane Co Ltd, and its operations moved to Hendon.

The largest school to move to Hendon in 1912 was the Ewen School. William Hugh Ewen originally founded his flying school at Lanark in 1911 before moving to Hendon. He was assisted by Sidney Pickles as chief instructor and Édouard Baumann as assistant, and obtained the sole agency for Caudron aircraft in Britain. W. H. Ewen Aviation Co Ltd was registered in January 1913, acquiring the business of Ewen and Andrew Mitchell Ramsay.

Breguet Aeroplanes Ltd was registered in 1912 with offices at 1 Albemarle Street; the company was run by R. Garnier and was the UK representative for the French parent company. It leased hangars at Hendon for the storage and operation of aircraft in this country, some of which were built under licence in its works at Willesden. By the end of 1912, therefore, Hendon was home to two British manufacturers (Grahame-White and Airco) and the agents of six French manufacturers (Bleriot, Breguet, Caudron, Deperdussin, Farman and Morane).

The first manager of the London Aerodrome had stated that there would be no 'gate money' meetings at Hendon to spoil the works of its tenants. Claude Grahame-White, however, had other ideas, and planned to hold events at Hendon every weekend and on Thursday evenings from Easter until the autumn. The First London Aviation Meeting, held over the Easter weekend of 5-8 April 1912, was the first event of the year. From then on the displays were held almost weekly until November. At that time there were no aircraft registrations except for those carried by military aircraft. In order to identify competitors, therefore, racing numbers were issued. Some were retained by the same pilots throughout their racing careers at Hendon, while others were often re-issued. Occasionally some pilots returned to Hendon after their number had been re-issued and had to be given a new one. Separate numbers were issued by the Royal Aero Club for major events.

The largest event of the year was the First Aerial Derby, organised by the Royal Aero Club and held on 8 June. The *Daily Mail* sponsored and promoted the race around London, which was won by TOM Sopwith. An estimated crowd of 45,000 spectators was at Hendon, and an estimated 3,000,000 around London witnessed the event. The day was ended by Grahame-White giving a demonstration of night-flying, a feat that had helped make him famous. The first entire display held at night took place on 26 September.

The legal action concerning the Wright Brothers' patents may have caused Grahame-White to lose interest in the United States, but he didn't lose interest in its people. It is claimed that he met Dorothy Cadwell Taylor, daughter of Bertrand L. Taylor of New York, on a transatlantic crossing, and in February 1912 they announced their engagement. On 27 June 1912 they were married at Widford church, with the reception at Hylands, the home of Sir Daniel Gooch, grandson of the railway engineer. The couple subsequently moved to Orange Hill House near Burnt Oak, the former home of the late Sir John Blundell Maple Bt.

The first display held at night took place on 26 September 1912 and ended with fireworks.

The only time Dorothy flew with Claude was when they returned from their honeymoon in France. It was not what she expected and terrified her, curing her of her desire for flying.

In July 1912 the company was sponsored by the *Daily Mail* to give seaplane demonstrations on the South Coast. Grahame-White planned to use a Curtiss Triad owned by Louis Paulhan, two Farmans and a Caudron. Frank Hucks was also employed for the flights, which were supported by a yacht carrying spares and other equipment. The *Daily Mail* also supported Salmet's exhibitions in the West of England, B. C. Hucks in the East Midlands and Gustav Hamel in the North East. Monsieur Fischer of the Farman company used one of the Farman seaplanes at Weymouth until joined by Monsieur Hubert when they headed to Exmouth.

In December 1912 the Wright Brothers' lawsuit was due to come to court and Claude's lawyers expected him to attend. He was unable to do so, having just returned from Spain in November, but his lawyers succeeded in getting the hearing postponed until January 1913. However, that month Grahame-White was in St Moritz with his wife enjoying bobsleighing, staying at the same hotel as Drexel and Santos Dumont. No more is heard in relation to the legal action against Claude Grahame-White, but in 1913 the court found against Glenn Curtiss. He continued to fight the decision in the courts and developed new aircraft, but the case stifled aircraft development in the USA. In 1913 the Wrights sued the representative of the British Government, Mervyn O'Gorman. In October 1914, with war having started, the action was settled for £15,000, removing the

TOM Sopwith, winner of the first aerial Derby in 1912.

threat of further litigation against military aircraft manufacturers in Britain.

The 1913 display season began earlier than the previous year. In February a speed contest was held in conjunction with the International Aero Show at Olympia. It was a handicap event run over heats and a final. The winner was E. Richer in a Breguet biplane, which caused alarm by banking steeply while turning around the pylons. The real start to the season came on the Easter weekend of 21-24 March when the Fourth London Aviation Meeting was held. After that almost every weekend was occupied, and some events had to be held on weekdays. With a full schedule of events, Grahame-White had no need to travel to satisfy his need for competition. Despite this he entered the first Schneider Trophy race at Monaco in 1913, but was refused permission to compete due to his late arrival. In the meantime he was able to concentrate on the promotion of Hendon and the management of his company.

Claude Grahame-White estimated that thirty race meetings had been held in 1912, with sixty trophies and £2,000 in prizes awarded for 100 contests watched by 500,000 visitors. For 1913 he estimated that fifty-one race meetings were held, with ninety trophies and £4,000 in prizes awarded for 200 contests watched by 1,000,000 visitors. I have found reference to thirty-three events in 1912 and fifty-four in 1913, not including the days set aside for exhibition and passenger-carrying flights. The season was a success despite the British holiday weather, which can always cause problems. The Second Aerial Derby of 1913, one of the highlights of the calendar, had to be postponed from June until September because of bad weather. The only other complaint was the amount of traffic in the area created by the displays at a time when few people owned cars.

Claude and Dorothy Grahame-White with their wedding reception guests at Hylands. In the background is Pierre Verrier's Farman Longhorn, and Grahame-White's Howard Wright is just out of shot to the right.

The Grahame-White Type VI Military Biplane, designed by John North.

The popularity of the events at Hendon led to complaints about the amount of traffic at a time when few could afford a motor car.

John Lawrence Hall and the Avro Type E with which he founded his school at Hendon.

During 1913 there were changes among the tenants at Hendon. The Temple School was renamed the Hall School of Flying when it was acquired by John Lawrence Hall. Hall was an engineer from Sheffield who had learned to fly at the Bleriot School, gaining his licence on 17 September 1912. The new school opened in August 1913 using an Avro biplane, but expanded quickly.

John William Dunne had been employed at The Balloon Factory before forming the Blair Atholl Aeroplane Syndicate Ltd to continue his research into stable aircraft. It was incorporated on 10 December 1909 with shareholders including his former colleagues Colonel John E. Capper and Alan D. Carden together with Frank McClean and the Duke of Westminster. In 1913 the company moved to Hendon from Eastchurch with Charles Richard Fairey as manager. On 8 May 1914 an exclusive licence was awarded to Sir W. G. Armstrong Whitworth & Co Ltd. Similar licences were awarded to Astra in France and Burgess in America, both of which constructed aircraft to Dunne's designs. The company was put into abeyance during the First World War, as was Nieuport (England) Ltd. This company had been registered in 1913 with temporary offices at 28 Milk Street, London. At first aircraft had to be imported and demonstrated at Hendon or Southampton, and George Miller Dyott was recruited as pilot. Aircraft production in England was planned but could not begin before the war started

While some companies enjoyed success, others struggled in 1913. Throughout the year Breguet Aeroplanes was under financial pressure and a receiver was appointed on two

The Dunne D.8 was an attempt to create an inherently stable aeroplane.

occasions until debts were cleared. It was a situation that could not be sustained and led to the company being wound up in 1914. British Deperdussin fared even worse after the founder of the French company, Armand Deperdussin, was arrested on charges of fraud. The managing director of the British company, D. Lawrence Santoni, faced similar charges in England relating to companies he had founded in his native country Italy, and both companies went into liquidation in 1913. Ironically both would eventually be acquired by Blériot, but the full story is outside the scope of this book.

In 1913 the Grahame-White factory produced the Bi-Rudder Bus, the forerunner of the company's successful Type XV. During that year the aircraft was used for the experimental fitting of a Lewis machine gun. The aircraft was tested at Bisley on 27 November 1913 with Marcus Manton flying it and Lt Stellingworf of the Belgian Army operating the gun from his seat between the undercarriage legs. Manton had qualified at the Grahame-White School and became an instructor and assistant to Louis Noel, who had succeeded Greswell as Chief Instructor at the Grahame-White School. The aeroplane may not have impressed the War Office but the machine gun did. In 1914 Birmingham Small Arms purchased the right to manufacture the gun and it entered service with the British Army in 1915.

It is impossible to know the influence that the Grahame-White company had on the aircraft industry. It is certainly the case that many who learned to fly at the school made a career in aviation, often being recruited by other aircraft manufacturers. Like many other engineering companies, it offered apprenticeships; one apprentice was W. D. (Doug) Hunter, who was at Hendon from 1913 until 1918. He went on to work with

The Grahame-White Type X Charabanc, one of the first passenger-carrying aircraft, used by Reginald Carr to break world records and win a Michelin Trophy

On 9 May 1914 William Newell made the first parachute jump from an aeroplane in England at London Aerodrome.

Kennedy, Vickers and Fairey before joining De Havilland at Stag Lane in 1925. He moved to De Havilland Canada in 1941 and eventually led the design team responsible for the Chipmunk, Beaver, Otter and Caribou before his death in 1961. Another apprentice was Frank T. Courtney, who became a renowned test pilot. Scholarships were also provided to assist those from poorer backgrounds to learn to fly.

There was no break between the 1913 and 1914 seasons. The last event of 1913 was the Christmas Meeting held on 25-28 December, and the first of 1914 was the New Year Meeting of 3 January. There were even more events planned for 1914 than in the previous year, but not all would be held. The displays were beginning to include aerobatics; this had started in 1913 with the demonstration of looping and inverted flying. Another new feature was parachuting. Balloons had been used for parachuting for many years, but on 9 May 1914 William Newell made the first parachute jump from an aeroplane during the May Meeting. *Flight* magazine praised the quality of the information published in the programmes for the various meetings. Such was the popularity of the events that grandstands were constructed, together with permanent cafes and tea rooms.

The Aerial Derby of 1914 was postponed because of persistent drizzle. The bad weather on that day also claimed the life of Gustav Hamel, who went missing over the English Channel while returning from France. When the new date for the Aerial Derby arrived the weather was almost as bad as it had been before, but despite this the race went ahead.

If you did not want to pay the entrance fee, the fields near St Mary's church provided an ideal vantage point to watch the displays. Land-owners charged 3d for access, which was half the price of entry to the airfield. In addition, refreshment could be obtained from the various public houses in Hendon, the Greyhound Inn being nearest. Hendon's management was less than happy with this arrangement, so screens were constructed near the railway embankment to obstruct the view from the top of the hill. It was impossible to stop people seeing all of the flying, but it was hoped that the screens would prevent them from seeing the start and finish of the races.

The bad weather affected the flying schools just as much as the air displays. Despite this, the number of pupils continued to grow and most schools began to employ more instructors. Lewis Turner at the Ewen School, however, was one who found himself alone when Ewen returned to Scotland. His departure saw the company change its name to British Caudron, referring to the aircraft the company marketed, and new instructors were recruited.

With space at Hendon limited and no room for expansion, Blériot could not begin aircraft production in England. Towards the end of 1913, therefore, he decided to move to Brooklands. A new factory was nearly ready in early May 1914 and soon aircraft construction could begin. Once the new factory was completed the school also moved to Brooklands, providing accommodation at Hendon for new tenants.

Airco, too, had planned to move aircraft production away from Hendon. A former ice rink on the High Street in Merton, Surrey, was acquired, but it is unclear whether the company moved there. What is known, however, is that a new factory was established next to the tram and bus depot in Colindale in 1914. The ice rink became the home of another of Holt Thomas's companies, Airships Ltd, which had E. T. Willows as its chief designer. In 1914 Airco was joined by the aircraft designer Geoffrey de Havilland, about whom much has

E. T. Willows erected an airship shed at Hendon in 1913, from where he operated his school until he joined Airships Ltd in October 1914.

been written. Unlike Blériot, however, this company retained its sheds at Hendon for the erection and test flying of aircraft.

The Beatty School was established at Hendon on 16 February 1914. George William Beatty was born in Whitehouse, New Jersey. He qualified as a pilot in 1911, and in 1912 began testing aircraft for the US Government at College Park, Maryland. In 1913 he brought his Wright Type B to England in the hope of marketing its Gyro rotary engine. He began teaching at Hendon and entered into an agreement with Handley Page; that company maintained the aircraft and he in turn instructed pupils of the Handley Page School. The London & Provincial Aviation Co also formed a flying School at Hendon in 1914. The founders of the company were William Thomas Warren, Michael Geoffrey and Walter Dorling Smiles.

In 1914 the Grahame-White factory and school still occupied the same row of sheds, although it had been extended slightly, but this was about to change. In November 1912

Handley Page had its factory at Cricklewood but used the London Aerodrome for all its flying.

William Thomas Warren. As well as founding a flying school he developed the Warren Helmet, one of the first items of protective flying clothing.

Grahame-White had told the press of his plans to fly across the Atlantic, and a British Trans-Atlantic Flight Fund was organised to raise money to cover the cost of the venture. The executive committee included Lord Lonsdale and Alfred Rothschild, while patrons included the writer Sir Arthur Conan Doyle, the explorer Sir Ernest Shackleton and the former Prime Minister Arthur Balfour. The challenge had been made even more attractive in 1913 by the offer of a £10,000 prize from the *Daily Mail.* The attempt had been planned for 1913 but it wasn't until 1914 that construction began of a hangar large enough to erect the aeroplane. The hangar was 200 by 100 feet, but unfortunately Grahame-White's ambitions were ahead of their time; in June 1914 the hangar collapsed before the roof was completed. Fortunately the accident didn't cost Grahame-White financially, but it did prevent a transatlantic flight before the start of the First World War. It is possible that the aircraft was designed by James Martin, who had also been raising funds for an attempt and expected construction to start in England in the winter of 1913/14.

The Maharanee of Tikari.

Claude Grahame-White was able to spend more time promoting the London Aerodrome and aviation in general during 1914. One often found royalty visiting Hendon, and many were given demonstration flights, often by Grahame-White himself. One of the more colourful members of the Indian aristocracy to visit Hendon was the Maharanee of Tikari, an Australian actress also known as Miss Elsie Forrest. She had married the Maharajah in 1909 after performing in India in 1904 and divorcing her former husband. With royalty including people like that, Grahame-White could be forgiven for being fooled when Lord Stanton Hope introduced him to the Crown Prince of Wurtemberg; the Lord was later revealed to be Robert W. Gunter, and the Crown Prince was a German clerk employed in the City of London.

The Ninth London Aviation Meeting was held over the bank holiday weekend of 1-3 August 1914, but it was marred by two events: bad weather typical of a British bank holiday, and a flying ban. On the last day the flying was curtailed when the expectation of war led to the imposition of a ban of flights more than 3 miles from an aerodrome. Before the ban was announced, however, Birchenough had taken a Maurice Farman to Leighton Buzzard for a flying demonstration, but he was able to return before evening. Henri Salmet was also able to return from a demonstration at Monmouth. He too had been unaware of the ban at the start of his flight and had intended to fly on to France. Pierre Verrier had also been away from Hendon, but was given Home Office permission to make the flight back from Coventry. Verrier, Salmet and the other French pilots now had to prepare to return home by train and ferry because of the ban.

A Southern Nigerian prince being shown the controls of a biplane at Hendon, circa 1913.

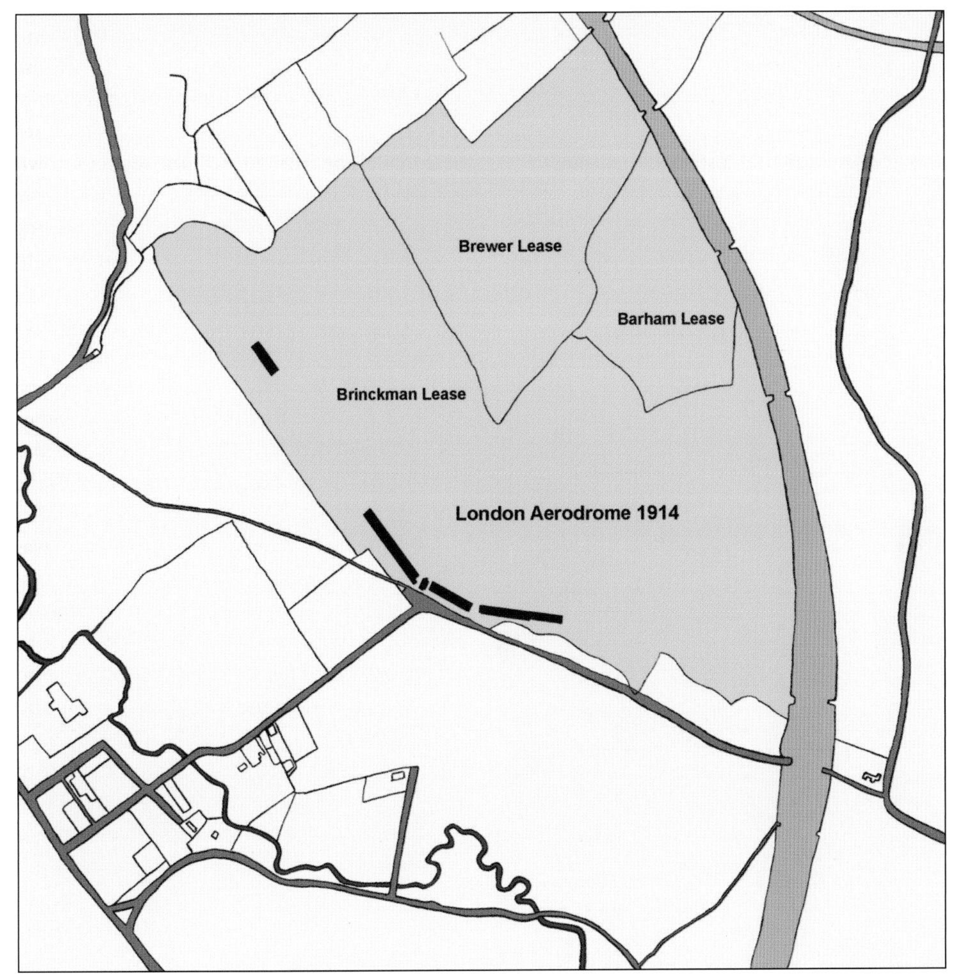

London Aerodrome, 1914.

3 The First World War

On 4 August 1914 Britain declared war on Germany, and on 8 August the last display was held. By then, however, many of Hendon's pilots had volunteered for active service with their countries' armed forces.

On the outbreak of war Louis Noel and the other French pilots at Hendon returned to France to serve in L'Aviation Militaire.

Louis Noel left the Grahame-White School and returned to France; his place was taken by Marcus Manton. Henry Charles Biard had qualified at Hendon in 1912; after brief service in the Royal Flying Corps he resigned his commission in 1914 and moved to France. When war broke out he returned from France and became an instructor at the Grahame-White School before joining the Royal Naval Air Service December 1917. It was fortunate that he did return because many others left Hendon and joined the RFC or RNAS.

Claude Grahame-White considered the First Lord of the Admiralty, Winston Churchill, to be a friend and immediately offered his services to the Royal Naval Air Service. On 7 August 1914 Murray Sueter, Director of the Air Department of the Admiralty, replied to Grahame-White's offer informing him of his intention to station aeroplanes at Hendon. Claude's company was to provide all assistance to the maintenance and operation of the aeroplanes and the production of spare parts. Claude Grahame-White was appointed Flight Commander, with Richard Gates and Eric Bentley Bauman appointed as Flight Lieutenants in the Royal Naval Reserve.

Part of the aerodrome was requisitioned by the Admiralty and John C. Porte was appointed Squadron Commander and placed in charge. He had been a naval officer before being invalided out and taking up flying. When the Deperdussin School had closed in 1913 he had moved to White & Thompson Co Ltd at Bognor, becoming involved with Curtiss flying boats; he was in America with Curtiss when the First World War started. He had been preparing to cross the Atlantic using a Curtiss America flying boat, but immediately returned to Britain, was accepted as an officer in the Royal Naval Air Service and became commanding officer of RNAS Hendon.

Contrary to the statement of Murray Sueter, no aircraft were delivered at first, most having been sent to France to provide air support for the Royal Navy and commercial shipping transferring the British Expeditionary Force to Belgium. Instead, many Hendon-based aircraft were impressed for military service and their owners compensated. The Eastchurch Naval Flying School was unable to continue training with most of its pilots in France so, once it became apparent that more pilots would be required, the Grahame-White School was recruited to provide elementary training.

Flying training wasn't the only role of the pilots still at Hendon because the Royal Navy had responsibility for the defence of the United Kingdom. The threat from airships bombing London had been recognised before the war started, so the pilots at Hendon were tasked with air defence patrols. On the night of 5/6 September Grahame-White and Richard Gates undertook the first night patrol over London as the result of a false alarm. Gates made another patrol on 14 September, but unfortunately crashed on landing and died of his injuries; he was buried in the churchyard of St Mary's,

Claude Grahame-White in the uniform of a Flight Commander in the RNAS.

RNAS aircraft at Dover preparing for the raid on Belgian towns in February 1915.

Hendon. The loss of the man who had done so much to develop Hendon was a setback for Grahame-White.

In early 1915 Claude Grahame-White revealed a different talent when he appeared as the star of a film. *The Secret of the Air* was made by the Trans-Atlantic Film Co Ltd and was planned for release on 21 January 1915. It made use of footage taken at Hendon and starred Claude as an airman. The film also included the late Gustav Hamel, but the plot wasn't revealed by the *Flight* correspondent, who saw an early version of it in December 1914.

The Cuxhaven raid of December 1914 was a pre-emptive strike at German airship sheds. Some authors claim Grahame-White took part, but this was not true. There was one occasion, however, when he did see operational flying, but it nearly cost him his life. In February 1915 the Royal Naval Air Service organised a major raid on German positions on the Belgian coast. Locations included Bruges, Zeebrugge, Ostend and Blankenberghe, and the raids were intended to prevent the establishment of U-boat bases in the area. Aircraft gathered at Dover and Dunkerque for the raid and pilots were brought from most air stations, including Hendon; John Porte, the Squadron Commander at Hendon, also took part. The official communiqué stated that thirty-four aircraft took part.

On 10 February the aircraft left Dover, but bad weather hampered operations and most were forced to land at Dunkerque; unfortunately Grahame-White was forced to ditch near Nieuport and was rescued by a French ship. Ray Sturtivant identifies the aircraft used as Farman Pusher Biplane 940. It is ironic that it should be this aircraft because this was the one Richard Gates was using when he had his fatal crash in September 1914. Following repair in the Grahame-White factory, it had been returned to service, only for it to try and claim a second victim. This is the last occasion when we hear of Grahame-White undertaking operational flying. On the following morning the others made another attempt, which was more successful. Claude was allowed to resign from the Royal Naval Air Service with effect from 29 July to concentrate on managing the company. Porte too left Hendon soon after, and on 16 July 1915 it was announced that Lt Cdr Christopher Hornby RN was appointed commander at Hendon.

The funeral of Richard Thomas Gates in St Mary's Church, Hendon.

Flying Training 4

The Admiralty had planned a complete takeover of Hendon and the eviction of the other flying schools. In December 1914 it relented, however, and the schools could continue operation, new buildings being erected for the use of the RNAS. By then, however, a new factory was under construction for Caudron in the former works of Morgan & Sharp, coachbuilders, at 255 Cricklewood Broadway. The company closed its school at Hendon and in January 1915 moved to its new factory. One of the school's instructors, Edwin T. Prosser, formed his own school using a single aeroplane. It barely lasted a month before the biplane was sold to Lawrence Hall and Prosser moved to Australia. The last civilian school to open at Hendon, the Ruffy-Baumann School, opened in 1915, taking its name from the Italian Félix Ruffy and the Swiss Édouard Baumann.

Conscription didn't start in Britain until 1916, so the flying schools could continue to operate much as they had before. Most of the initial pilot training for the RNAS and RFC was undertaken at civilian schools. There were also some individuals who still wished to learn to fly and sought private tuition. Many of the replacement flying instructors were recruited from the best of these.

Édouard Baumann, seen here, had been employed at the Beatty School as assistant instructor, and Félix Ruffy was one of his best pupils.

Pupils of the Ruffy-Baumann School waiting their turn 1915.

In 1915 a new Naval Flying School was formed. Initially it was part of the Grahame-White School, but gradually it established its own identity under its civilian instructor, Frederick Walter Merriam, the only one in the RNAS. He had been an instructor at the Bristol School, Brooklands, and when war started he offered his services to the RNAS just as Grahame-White had done. Initially he was rejected because of his poor eyesight, but was subsequently told to report to Hendon where he became the chief instructor of the Naval Flying School. His duties were various, including the testing of aircraft and pupils. Different pupils from the various schools displayed varying ability and aptitude. Merriam reckoned the best were those trained on pusher aircraft at the Grahame-White School under the supervision of Marcus Manton; those trained on tractor aircraft were among the worst. No consideration had been given either as to whether the newly qualified pilots would be suitable as officers in the RNAS.

Marcus Manton making an entry in his logbook in 1916. From December 1914 the pupils and instructors of the Grahame-White School used this former London bus as a shelter while awaiting their turn to fly.

The skies around Hendon were getting very busy, so the location for a sub-station was sought. The reasons are unclear, but Chingford was chosen and construction of the airfield begun. In March 1915 it was also selected to be a night emergency landing ground and home to a defence flight. Chelmsford was selected as the location of another home defence flight administered by Hendon. Naval Air Service Confidential General Memorandum No 12 was issued by the Air Department on 9 April 1915 listing Chingford and Chelmsford as sub-stations of Hendon. Chingford officially opened at the beginning of May with Air Mechanic Reggie Webb being one of the first twelve mechanics to transfer on 1 May 1915.

When the Naval Flying School was opened at Chingford many of the instructors and aircraft were taken from Hendon. By 24 May 1915 Squadron Commander John C. Porte and the other officers were listed under Chingford, indicating that it had taken over as the home of the Naval Flying School, despite the fact there were more pilots under training at

Instructors and pupils at the Grahame-White School.

Hendon. Gradually, however, the numbers of pupils at Chingford grew until the school at Hendon was closed. Once the Admiralty work was finished the Grahame-White School opened its doors to all pupils.

The anti-airships patrols of 1914 had been in response to false alarms. The first actual raid on London took place on the night of 31 May/1 June 1915, causing damage to the East End. On the night of 8/9 September London suffered one of its worst attacks from a single airship. The first bombs were dropped on Golders Green at about 2240 hours but most fell between Euston and Liverpool Street stations, killing twenty-six, injuring ninety-four and causing more than £500,000 worth of damage. These raids were not the first on British soil, but they caused greater public anger at the state of Britain's air defence.

On 16 October 1914 the Admiralty had assumed control for the air defence of London, and the War Office for the defence of military targets. The flying instructors at the Naval Flying School took over as pilots for the anti-Zeppelin patrols previously flown by the Grahame-White School. Even Frederick

A Sopwith Gunbus. One was used to try and intercept the Zeppelin raid of 31 May 1915, but it crashed, killing Flight Lieutenant Douglas Meston Barnes and injuring Flight Sub-Lieutenant Benjamin Travers.

A new shed was erected to replace the original Aeronautical Syndicate and Bleriot sheds, wrecked by a snowstorm in March 1916.

Warren Merriam, while still officially a civilian flying instructor, flew some of the patrols from Chingford. Airship raids proved difficult to track and intercept. Aircraft were supplied from whatever was available and the RFC was instructed to assist where possible; 17 Reserve Squadron flew at least one sortie from Hendon in January 1916.

The author and playwright George Bernard Shaw, one of many visitors to Hendon during the First World War.

There was a continuous discussion about the best way to organise defences; decisions could change rapidly. Finally it was decided that the Navy would be responsible for defence over the sea and the Army would be responsible for defence over land. Admiral Sir Percy Scott had been responsible for the defence of London, but on 16 February 1916 this passed to Field-Marshal Sir John French, C-in-C Home Forces. RNAS air defence night flying at Chingford and Hendon was stopped and the personnel instructed to proceed to Yarmouth with their aircraft. Their place was taken by 19 Reserve Squadron RFC. This was formed at Hounslow on 29 January 1916 with home defence detachments of two aircraft each being sent to various aerodromes, including Hendon. On 15 April the squadron was renamed 39 (Home Defence) Squadron, but by 1 May 1916 it no longer used Hendon as it concentrated its aircraft at Suttons Farm and Hainault Farm.

Questions had been raised about the observation of RAeC certificate tests in August 1915. The Assistant Secretary of the Royal Aero Club met with Bernard Isaac at Hendon and discussed a number of issues. It was proposed that letters be sent to the schools and the observers reminding them of their responsibilities. It appears that the problem wasn't resolved because later documents imply that the RFC was still dissatisfied with the standard of instruction at the civilian flying schools. A conference was held on 20 June 1916 to discuss the matter, at which seven schools were represented. The RFC proposed to raise the fee from £75 to £125 for officers gaining the RAeC certificate. In return the schools were to abide by regulations laid down by the RAeC and be supervised and inspected by officers of the RFC. The Air Board agreed to the findings of the conference on 26 June and that the schools should be given another chance; if found useless they would be taken over by the War Office.

On 12 August 1916 a letter was sent from the Director of Air Organisation (DAO) to ten schools, including all those at Hendon, laying down the conditions under which the schools would operate and be governed. After the original ten had received the letter others applied for and accepted the agreement. In April 1917 the Cheltenham & West of England Aviation Co and Wells Aviation Co had their contracts cancelled, followed in May by the Cambridge School of Aviation. The Cheltenham and Cambridge schools had been registered by Ruffy in 1916 and little more is known of either school. Wells Aviation was formed in 1916 in Chelsea and recruited Sydney Pickles as its chief instructor. He resigned, however, in March 1917 and the company went into liquidation in May 1917.

The Hendon schools that accepted the contract to provide initial flying training were the Grahame-White School, London & Provincial, Ruffy-Baumann and Beatty. These were formed into the RFC School of Instruction, which came under 18 Wing on 22 September 1916. A Commandant was appointed, Major Leslie Fitzroy Richard, whose main task was the allocation of pupils to the schools, each one with a junior RFC officer in charge. H. P. G. Brackenridge, manager of the Grahame-White School, often wrote requesting more pupils. Even when he did get them he had to write once more requesting full details. The school instructors were exempt from conscription, but were Class W Reservists. Despite this one of the frequent tasks of the Commandant was correspondence with the appropriate authorities to ensure this did not happen. If an employee was dismissed from one school he also ensured that he wasn't taken on by one of the others, and that he was notified to the authorities as eligible for conscription.

The Hall School was one of the few that didn't become part of the RFC School of Instruction. The date of its closure is uncertain, but was probably in the spring of 1917; Lawrence Hall's activities until his death on Christmas Day 1920 are unknown. Cecil

Pupils and instructors of the Grahame-White School after it had become part of the RFC School of Instruction.

The Hall School.

McKenzie Hill, however, who had been an instructor at the school, moved to New Zealand in 1917 and trained more than 100 pilots before being killed in a flying accident on 1 February 1919.

The probable reason for the closure of the Hall School was its exclusion from the aerodrome in the early part of 1917, when all the schools were given notice to quit. The

Staff of the Beatty School after their move to Cricklewood.

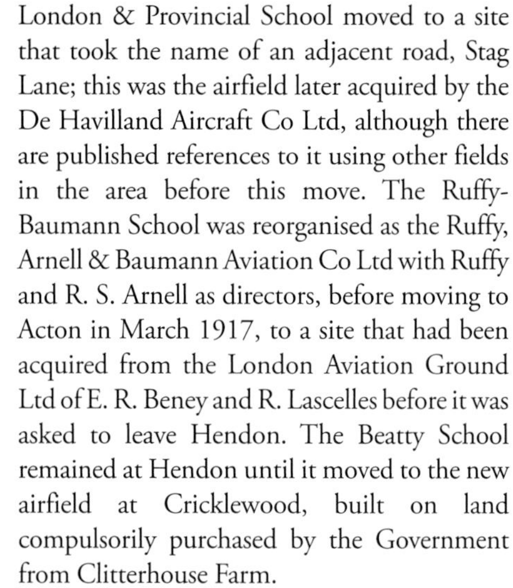

London & Provincial School moved to a site that took the name of an adjacent road, Stag Lane; this was the airfield later acquired by the De Havilland Aircraft Co Ltd, although there are published references to it using other fields in the area before this move. The Ruffy-Baumann School was reorganised as the Ruffy, Arnell & Baumann Aviation Co Ltd with Ruffy and R. S. Arnell as directors, before moving to Acton in March 1917, to a site that had been acquired from the London Aviation Ground Ltd of E. R. Beney and R. Lascelles before it was asked to leave Hendon. The Beatty School remained at Hendon until it moved to the new airfield at Cricklewood, built on land compulsorily purchased by the Government from Clitterhouse Farm.

The Grahame-White School moved across Aerodrome Road onto 'Little Hendon', which it leased for 99 years from William Lyulph Johnson on 15 March 1917. This was far from ideal, however, and on 2 July 1917 OC RFC School of Instruction wrote to OC 18 Wing requesting the use of Hendon Aerodrome between 4.15am and 7.30am by pupils making their first few solo flights. He pointed out that the school's new aerodrome was small and had difficult approaches, and that the main aerodrome was used exclusively by the

AAP and to date the school had not been allowed to use it. Permission was granted on 13 July, but would have been ended as dawn became later.

The Grahame-White instructors continued to give exhibitions at weekends, attracting crowds of spectators who wandered onto the aerodrome, just as they had before the war. One unfavourable report identified that there was no control over people entering the aerodrome. On 31 July 1917 Major Richard had to ask Brackenridge to prevent it happening on the school aerodrome. On 29 October the flying instructors at the Grahame-White School, Meering, then Pashley, Biard, Winter and Osipenko, were made available for conscription, but next day they were reinstated. No reasons are given, but this was a period of industrial unrest within the aircraft industry. Meering was subsequently charged with theft of company property but acquitted on 11 March 1918, the Chairman of the Magistrates apologising to him for having to undergo a trial.

On 5 September 1917 a conference was held that recommended an increase in the fees paid to the schools. Once the increase had been approved the Treasury wrote to the four Hendon schools on 22 December offering an increase of £135 per pupil and a £25 bonus to take effect from 1 October 1917. A new letter was issued by the DAO on 11 January confirming the new arrangements, but bringing them into effect from 22 December 1917. These badly worded documents were to form the basis of one of Grahame-White's claims on the Air Ministry; the other schools had similar issues. Legal opinion in 1920 supported the schools' claims that the revised figure applied to all who qualified after 1 October 1917 and not just those who had joined the schools after that date.

In December 1917 there were 126 pilots under training, with most at the Grahame-White School. Billets were found for them in houses in the area and they had their mess at Hendon. Roll-call took place at the mess, then transport was provided to get them to the various aerodromes. The vehicles available to the School of Instruction consisted of two Leyland 3-ton heavy tenders, one Crossley 15cwt light tender, one Ford car, three Ford vans, a Ford ambulance and two P&M motor cycles with sidecars.

The contract with the Beatty School was cancelled in early 1918 and it closed on 5 February. On the 24th HQ Training Division issued an unfavourable report on the remaining Hendon schools, advising that they should be abolished or reconstructed. On 22 June 1918 the remaining schools, including Grahame-White, were notified that their contracts would end with effect from 1 July 1918 '...in consequence of the changes which have taken place in the training requirements now demanded...'.

Aircraft of the Grahame-White School at 'Little Hendon'.

5 Aircraft Acceptance

When the Aeronautical Inspection Department (AID) was formed it was intended to inspect aircraft and equipment for both the Military and Naval Wings of the RFC. When the Naval Wing became the Royal Naval Air Service it lost access to AID and had to establish its own organisation. This meant that an additional task for the flying instructors at Hendon was that of test pilot.

Hendon became a reception depot for aircraft built for the RNAS, resulting in the construction of new buildings in the south-east

The Burgess Gunbus. The RNAS ordered aircraft from companies in North America, most of which were assembled and test flown at Hendon.

corner of the aerodrome. Aircraft that were delivered by road needed assembling and all needed test flying, often by the staff of the Naval Flying School. When the school moved to Chingford in 1915 some RNAS personnel remained at Hendon under Flight Commander C. Hornby, sharing the task with Chingford.

Many of the aircraft received at Hendon came from North America, where a purchasing commission had placed orders with Burgess, Curtiss, Sloane-Day and Thomas in the United States and Curtiss in Canada. Deliveries began in 1915 and continued until 1917. Many of the later batches of aircraft, however, were delivered straight to store or broken up on their arrival in Britain. Once these orders were completed Hendon assumed a more general role, receiving aircraft from London factories.

One of the most disruptive aircraft deliveries took place on 9 December 1915. Handley Page of Cricklewood had received an order for the O/100, the first large bomber built in England. It was too big for the existing factory so Handley Page acquired extra space near the Airco works in Colindale. The prototype had to be taken from the factory by towing it along Colindale Avenue after the street furniture had been removed and overhanging trees were brutally pruned. Fortunately the aircraft arrived unscathed and made its maiden flight on 17 December; the settlement of claims for damages took longer to resolve.

Until 1915 all RFC aircraft were taken to Farnborough for inspection and acceptance. With increasingly large orders the system could not cope, so inspectors were placed in the larger factories, and facilities similar to those at Farnborough were opened at Hendon. These in turn were supplemented by others around the country to which the manufacturer delivered the aircraft. If aircraft arrived by road they were erected before being test flown by a RFC pilot seconded to AID. An officer was placed in charge of the AID aerodrome at Hendon and supervised the work.

In 1916 pressure from within Parliament and public opinion prompted the Government to form a Committee of Inquiry chaired by Mr Justice Bailhache. In November 1916 the Committee recommended in its report the establishment of an Air Board with powers to supervise the design, construction and supply of aeroplanes to the two Services. It was constituted as a ministry under the Ministries & Secretaries Act 1916 and its first meeting was held on 3 January 1917. Responsibility for aircraft manufacture, however, was transferred to the Ministry of Munitions.

The transfer of aircraft production to the Ministry of Munitions had a significant effect on Hendon. AID was transferred to the Ministry of Munitions, so control of its former aerodromes was passed to the Directorate of Aircraft Equipment. When this happened the remainder of the aerodrome was requisitioned,

The second Handley Page O/100 after delivery to the RNAS at Hendon, showing how difficult it was to manhandle large aircraft on the ground.

a fence was erected to separate it from the factory, recreational flying was stopped, and the flying schools were given notice to quit. *Flight* magazine for 8 February 1917 regretted the loss of public access to Hendon, but looked forward to the day when enthusiasts could return.

Additional land was requisitioned under the Defence of the Realm Act for the construction of a railway around the airfield to serve Hendon and the factories on the Edgware Road. Approval for its construction was given on 18 April 1918 and it was completed as far as the factory platform by 19 May 1918. It was built by the Midland Railway and ran north from the main line, crossed Aerodrome Road and ran around the northern perimeter of the airfield. It then turned south and a spur ran behind the sheds of the Aircraft Acceptance Park in the south-west corner of the airfield. The rest of the line continued south-west and stopped east of Edgware Road just north of the tram depot, providing rail access to the Airco factory and Hendon Aircraft Production Depot. The locomotive shed and bridge across the Silkstream can still be seen in Montrose Park.

The Director of Aircraft Equipment (reporting to the Director-General of Military Aeronautics at the War Office) and the Superintendent of Aircraft Construction (reporting to the Director of Air Services at the Admiralty) came under the control of the Air Board in 1917. The former AID aerodromes became Acceptance Parks and the RNAS acceptance system was gradually absorbed. Hendon Aircraft Acceptance Park was formed on 22 March 1917 with Major Lord Robert Innes-Ker as its first Park Commandant. The other officers consisted of the Adjutant & Dispatch Officer, Equipment Officer, and test and delivery pilots; two AID officials were also present. While it received RFC aircraft in the west, the RNAS aircraft continued to be received in the east. Lt A. Scott RNAS had been an inspector at the Grahame-White factory between 1915 and March 1917, but by the time King George V visited the factory in December 1917 Lt Reid RFC was Production Officer and John S. C. Henderson the Chief AID Inspector.

From 14 August 1916 Flt Lt Theodore Douglas Hallam DSC had been senior naval officer at Hendon, succeeding Flt Cdr Henry Richard Busteed, to whom he had been assistant. Hallam noted in his autobiography that 'the work at Hendon was petering out, the soldiers of the RFC had cast a monocled and covetous eye on the aerodrome' and decided to leave. He was posted to Felixstowe and by 12 March 1917 Wg Cdr Charles Erskine Risk was the senior officer. On 16 April 1917 Squadron Commander Francis George Brodribb took command of the RNAS at Hendon, but when King George V visited on 31 May it was Innes-Ker who received him, suggesting that he was in overall command and had been the one to evict the flying Schools. Brodribb subsequently left on 28 March 1918 and became the first commanding officer of 1 (Hamble) Marine Acceptance Depot.

Before the Royal visit of May an experimental catapult had been erected on the airfield, north of the flying ground. It was designed by R. F. Carey, built by Waygood-Otis and used compressed air. The first launch of an Avro 504H was made with Flight Commander Rupert E. Penny at the controls, and the aircraft used could have been N5361 or N5270, both of which were at Hendon between 22 and 25 May 1917. N5269 had also been allocated for use in the trials. Once the tests were completed the catapult was removed.

The daylight aeroplane raids of 1917 saw Hendon briefly assume an air defence role once more. Aircraft Acceptance Parks were expected to have a number of modern aircraft at any one time and the Air Board decided that six DH.4 aircraft should be maintained for the defence of London. This could not always be maintained, however, because of the need to dispatch them to front-line squadrons, and there were occasions when the pilots used anything that was available. Despite this a number of sorties were flown, mainly using Airco DH.4 and DH.5 aircraft in June and July. At least the raids encouraged the move of the pilots from tents on the aerodrome to billets in Colindale Avenue.

In 1917 various aircraft found their way to Hendon. Most were provided for the use of officers based in London for either flying practice or travelling. They varied from the elderly BE.2c to the latest Sopwith Snipe, and were generally referred to as belonging to the AAP or the station, but can be considered as forming a Communications Flight.

In September 1917 Hendon AAP had 435 personnel and erected 130 aircraft, the most for

This was the result of an attempt to intercept the German daylight raid on London on 7 July 1917.

any of the nine Acceptance Parks formed, though Lympne had more personnel. On 12 October 1917 the designation changed to 2 (Hendon) AAP, and in December its three sections managed only twenty-seven per week (about 108 per month) out of a possible seventy-five per week (twenty-five per section), despite an increase in personnel to 445. The figure was lower than it had been, but was still higher than any AAP. The problem was that they were below established strength and aircraft were not received from the manufacturers quickly enough.

When the Royal Air Force was formed on 1 April 1918 there was no longer a separate RNAS or RFC. In addition, a change of organisation placed the AAP in 6 (Equipment) Group of South East Area. By July 1918 a Handley Page Section had been added to the AAP, using the three Belfast truss hangars beside the railway line. The Communications Flight increased in size and became 1 (Communications) Squadron on 26 July 1918. A Medical Flight had also formed at Hendon in 1918, tasked with returning pilots to flying duties. It was redesignated 29 Training Squadron on 1 July before moving with its DH.4 aircraft to the London & Provincial School's airfield at Stag Lane on 6 August.

At the Armistice the AID locally was organised into the Hendon District which was split into the Hendon Group which covered Airco, Grahame-White, Kingsbury, Integral and London & Provincial, the Willesden Group,

A DH.4 converted to carry passengers and used by the Acceptance Park Communications Flight.

An Airco DH.6 built by Grahame-White Aviation at Hendon in about July 1917; the Belfast truss hangars can be seen under construction in the background

the Cricklewood Group and the Hayes Group. There was also the Hendon Aircraft Production Depot and the AGS Bond, West Hendon.

The Hendon Aircraft Production Depot opened in 1918 at The Hyde, between the Airco works and Stag Lane aerodrome. Contrary to its name it was actually responsible for the recovery of reusable aircraft components and was originally to have been called 1 Aircraft Salvage Depot. It also serviced some of the fleet of vehicles used by the Directorate of Aircraft Production. It was redesignated 1 National Aircraft (Salvage) Depot on 9 January 1919 and closed on 1 September 1919. The AGS Bond was responsible for the receipt and inspection of components manufactured by companies too small for a permanent inspector to be appointed. By reducing the time inspectors spent travelling they could spend more on inspection.

By November 1918 control of 2 (Hendon) AAP had passed to 1 Group following a reorganisation. An Air Ministry Order of December 1918 instructed pilots visiting London not to land at Hendon but to proceed to Hounslow, where facilities were available to receive them. One must assume, therefore, that facilities were being reduced quite quickly. The AAP officially closed on 10 January 1919, but despite this a Minute to the Director of

The Hendon Aircraft Production Depot, still showing its original designation of No 1 Aircraft Salvage Depot. It also serviced some of the fleet of vehicles used by the Directorate of Aircraft Production.

Equipment dated 19 May stated that Hendon was working to clear the park of large machines and during the previous month had only produced three DH.9As for the Expeditionary Force. A file on the 'Future Policy of Aircraft Acceptance Parks' anticipated the following deliveries from Hendon in May-June 1919:

Airco DH.9A	3 per week for Expeditionary Force – machines in Park
Airco DH.10	4 per week
Nieuport Nighthawk	1st machine off new contract for test
(Caudron) Nighthawk	1st machine off new contract for test
(Morgan) Vimy	2 per week if modification embodied

Flight commented that aircraft were still being received at Hendon in August 1919 at a rate similar to those leaving.

When 86 Wing was formed on 13 December 1918 for communications with the Paris Peace Delegation, control of 1 (Communications) Squadron was transferred to it from the AAP. Both the wing and the squadron moved to Kenley on 17 April 1919. At some point the acceptance work also came to an end. A memo titled 'Policy for Civil Aerodromes' by the Controller General of Civil Aviation dated 30 June 1919 includes an appendix of stations for disposal and therefore available for civilian use; Hendon is listed in this appendix with the exception of ten sheds that were probably still being used for the storage of RAF aircraft.

In order to reduce the large stock of relatively new aircraft at RAF stations around the country the British Government gave some as gifts to countries in recognition of the support they had provided in wartime. At a ceremony at Hendon in February 1919 fifteen aircraft were presented to the Canadian military authorities, part of a larger gift so that they could form their own flying corps. Having prepared them for the ceremony, the remaining staff of the AAP probably had to crate them ready for shipment to Canada.

The Disposals Board of the Ministry of Munitions sold off war surplus material, and in June 1919 some of the material for disposal at 1 National Aircraft (Salvage) Depot could be viewed at the Agricultural Halls, Islington. Before then, however, an auction of motor vehicles was held at Hendon in April, followed in May by one for aircraft and materials. Although another sale of vehicles was planned for November, no more auctions appear to have been held before the remaining stock was sold to the Aircraft Disposal Co Ltd in early 1920.

Once Hendon had been cleared of remaining aircraft and other RAF property it was intended that it should be returned to its owners. The final clearance was the responsibility of a Care and Maintenance Party from RAF Uxbridge, but in the spring of 1920 their work too was completed.

O/400 bombers can still be seen in store in the background of this view of the Victory Derby of 1919.

6 Aircraft Production

The Hendon schools were unusual in that all of them undertook aircraft construction. London & Provincial employed the former Martinsyde designer Anthony A. Fletcher. New works were acquired in 1916 in the former London & Parisian Motor Garage on Colindale Avenue, before moving to the former British Caudron factory works in Cricklewood. The Ruffy-Baumann School had built a number of Caudron copies for its school, while the Hall School used Caudron wings fitted to a fuselage resembling that of an Avro Type E.

The Beatty School had used Wright Type B biplanes built under licence and known as Beatty-Wrights. The factory at Cricklewood was in use from 1916 and was used to construct the Beatty Biplane. From 5 November 1917 until 14 April 1918 a number of components were ordered from the company; most were parts for Handley Page aircraft, but there was also an order for 100 mainplanes for Airco. The company ceased trading at the end of November 1919.

Hendon also attracted other companies seeking a suitable location to test fly their designs. The Mann & Grimmer biplane was an attempt at a new arrangement of warplane. It had a single engine driving two propellers, which gave an excellent field of fire for a gunner but was over-complicated. It was built in Surbiton and tested at Hendon in 1915, first using one of the London & Provincial sheds before moving into one of the Admiralty sheds. It returned to Hendon after a rebuild but no hangar space was available so the firm had to erect a tent near the Hall School. In August 1915 the aeroplane was grounded by the lack of a test pilot. In October A. E. Barrs took over test flying, during which it suffered from a number of problems with its transmission. The last transmission failure resulted in the machine being wrecked on 16 November 1915; an improved replacement was not completed before backing was withdrawn.

Late in 1916 the company F. Nestler Ltd experimented with aircraft construction. Previously it had an agency for Sanchez-Besa aircraft, one of which was demonstrated at Hendon in May 1914, and was a wartime sub-contractor. A single-seat scout was built to the design of Monsieur E. Boudot and transported to Hendon for evaluation by the RNAS. It was test flown by J. Bernard Fitzsimmons in a manner that impressed observers. Unfortunately a wing failed and he was killed when it crashed into the roof of one of the London & Provincial School hangars. Fitzsimmons had qualified in 1915 at the Ruffy-Baumann School before becoming a freelance test pilot. He was an instructor at the Grahame-White School before joining Nestler in March 1917.

Unlike the Mann & Grimmer biplane, the Nestler Scout had shown promise. Even if it had been a success, its adoption by the Air Services would have been a lottery. An advocate would have had to convince the Admiralty or War Office of the value of the aeroplane to them, and success would have depended to a great extent on the advocate's personality. If successful orders would only be forthcoming if they could be fulfilled; therefore the company needed production facilities, which it lacked. From 1 April 1917 the Minister of Munitions banned 'the experimental manufacture of any aeroplane or seaplane or any part thereof' without a licence from the Ministry. The ban excluded those built to Government orders but effectively prevented further projects such as those undertaken by Mann & Grimmer and Nestler. It also brought to an end the manufacture of aircraft by the Hendon flying schools, with the exception of Beatty and Grahame-White.

Claude Grahame-White's personal life suffered at the start of the First World War. His

A wartime Beatty-Wright biplane. They were used exclusively by the Beatty School.

The Mann & Grimmer biplane was hampered by its complicated engine and propeller arrangement, one abandoned by Beatty.

The Nestler Scout in front of the Grahame-White factory.

wife Dorothy had sailed to New York in 11 November 1914, and when she returned to England she petitioned for the restoration of conjugal rights followed by divorce proceedings. Newspapers at the time referred to an 'American actress' being named in the case and the divorce was granted on 7 June 1916. Claude married his second wife, Ethel Grace Levey, the former wife of George M. Cohen, at St Marylebone Registry Office on 21 December 1916, and they moved to Piggots Manor, Letchmore Heath. She was probably the actress named in the divorce proceedings. Claude may have met her as early as May 1913 when he took her for a flight at the Empire Day Meeting; this was the first time her name was mentioned in connection with Hendon. She had custody of her daughter from her previous marriage, Georgia Ethelia (Georgette) Cohen, who retained her father's name but now spent most of her time in England.

Dorothy C. Taylor reverted to her maiden name before marrying the Count Carlo Dentice di Frasso in 1923.

While there was all this disruption in his private life, Claude still had to attend to his business. Frederick Harrold Payne was taken on as a director in place of the late Richard Gates, and he became joint managing director, removing some of the work from Grahame-White. When the Society of British Aircraft Constructors was formed in 1916, Payne represented the company at its meetings. There were clauses in the original property leases for a review after five years. In 1915 Grahame-White took the opportunity to sign new 99-year leases, acquiring in the process an additional 13 acres near Colindale Avenue from Church Farm and an additional 26 acres of Featherstone Farm.

The Grahame-White company was seeking larger orders to justify factory expansion. The first were in batches generally of twelve or twenty-four aircraft. The first wartime order was for twelve Morane-Saulnier Type H monoplanes for the RFC, followed by a further batch of twenty-four. The first production aircraft for the RNAS was for BE.2c aircraft; a batch of twenty-four was delivered between July 1915 and January 1916 at the rate of four per month. A further batch of twelve was delivered between April and July 1916 at the rate of three per month. A far higher

A Morane-Saulnier Type H, one of the last of thirty-six built for the RFC.

production rate was achieved for the Grahame-White Type XV, also known as the Admiralty Type 1600, the prototype of which was delivered in May 1915. Thirty-four production aircraft were delivered between August 1915 and February 1916 at a rate of five per month, with most allocated to Chingford and the Grahame-White School.

The Admiralty had a number of issues with the company. Its complaints of 1915 resembled those of the War Office in 1912, including slow delivery rates. In return, however, Grahame-White had a number of complaints about the Admiralty, as revealed in a letter dated 3 May 1915. Winston Churchill had visited the factory at Hendon in April 1915 and asked Grahame-White what further orders the company could cope with. When Grahame-White offered to double the size of the factory in return for an order of 100 aircraft, J. E. Masterton Smith raised a number of issues with him on behalf of Churchill. In return Grahame-White pointed out the many problems caused by the Admiralty, not least the problem of bidding for a contract to build BE.2c aircraft based on the work they had done building two BE.2a aircraft; the two are very different in design. Insufficient drawings were provided

Type H monoplanes under construction in the Grahame-White factory.

First Lord of the Admiralty Winston Churchill at Hendon, inspecting a BE.2c built by Grahame-White Aviation.

and it was discovered that the materials specified for the BE.2c were different from the BE.2a. The company was then required to incorporate modifications once the aircraft were almost complete.

Grahame-White faced a chicken-and-egg situation. The Admiralty was not prepared to increase the size of the orders unless the factory was large enough to cope. Grahame-White needed those orders, however, to justify the cost of expansion. Eventually new factory buildings were erected and production increased, paid for in part by the Admiralty. At the same time more workers were recruited; the number of employees had grown from about twenty-four in July 1911 to more than 400 in 1916. The Military Service Act of 1916 made it harder to recruit skilled workers, and those that remained could seek the best-paid jobs. Grahame-White had no choice but to recruit unskilled men and women and train them, hoping that they would not be lured away by better wages elsewhere, especially at Airco and Handley Page.

When Grahame-White expanded the factory it allowed the authorities to place larger orders, and 100 Grahame-White Type XVs were built in two batches of fifty, one for the RFC and the other for the RNAS. By the time

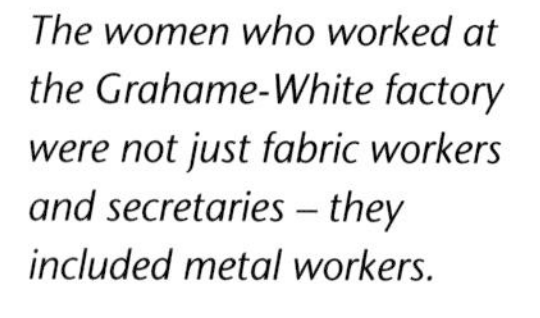

The women who worked at the Grahame-White factory were not just fabric workers and secretaries – they included metal workers.

The Grahame-White Type XXI was a contemporary of the Nestler Scout.

the orders were completed, however, the type was obsolete and many were delivered to store; some were subsequently returned to Grahame-White in 1917 and may have been used by the school. These aircraft were followed by two batches, each of 100 Henry Farman F.20 aircraft. The orders were originally placed with the Aircraft Manufacturing Co but sub-let to Grahame-White.

John D. North had left Hendon in 1915 and moved to the Aviation Department of the Austin Motor Co in Birmingham. Despite this the Grahame-White factory produced new types after his departure. The Grahame-White Type XIX was in fact a licence-built version of the Breguet V. Initially thirty were ordered in 1916, but in November of that year the final twenty were cancelled. The first was not delivered until January 1917 and the remaining nine may have been delivered straight to store. In 1916 the Type XX two-seat trainer was built, followed in 1917 by the Type XXI single-seat scout; neither was registered or put into production. Monsieur E. Boudot left Nestler and joined Grahame-White, where he designed the Grahame-White E.IV Ganymede. Three prototypes were ordered in 1917, but only one was completed.

It was not just the production facilities that needed expansion. In March 1916 Grahame-White opened new mess rooms for the company workers, managed by the YMCA. At the same time management of the cafes in the enclosures was taken in-house. The Paddock Café and Bar had also been refurbished and reopened in March 1916. Such was the need for skilled labour that it was difficult to find housing for them in an area that was still quite rural. Aeroville was built by the Grahame-White company to house some of its employees, and was the only part of a large estate planned by Herbert W. Matthews that was actually built. Even in 1919 they hoped to continue the development. Ironically they would have had electricity and modern appliances despite being based on 18th-century domestic architecture, the whole estate resembling the squares of the Inns of Court in London.

It was not all work and no play for the factory workers. In June 1916 the inaugural Aircraft Workers Sports Day was held on Hendon Aerodrome under the auspices of the YMCA. Various races and other events were held on the airfield for both men and women, despite an attempt by the rain to spoil the day. A second Aircraft Workers Sports Day was held in July 1917 at Stamford Bridge, but subsequently each factory organised its own sports days. The Grahame-White Recreation Association held its first sports day on 27 July 1918, and another was held on 24 August.

In 1917 there was an exodus of officers from the company. Furniture manufacturers in High Wycombe had undertaken aircraft sub-contract work for Airco and Handley Page, but complained about the disruption it caused. To remedy this George Holt Thomas founded Wycombe Aircraft Constructors Ltd. Frederick H. Payne was recruited as managing director of the new company from Grahame-White, and took with him Thomas E. Ritchie as general manager and T. Kemp Walton as company secretary. The war ended before the factory was completed and the company was wound up in 1923.

Further expansion of the factory took place in 1917, allowing Grahame-White to build 750 Airco DH.6 and 900 Avro 504K aircraft, most of which were used as training aircraft. This

One of the assembly shops, with Farman F.20 nacelles in the foreground and Airco DH.6 aircraft beyond.

expansion was funded by the Ministry of Munitions, but ironically it was this support that would be the cause of the post-war problems. In December 1917 Stanley Baldwin was asked to approve an extended loan to Grahame-White of £320,000. The company only had £15,000 of capital but liabilities of £130,000. Orders worth £700,000 had been placed with the company and it had already received a loan of £70,000, which would be incorporated into the new loan, effectively providing £250,000 of new money for the company. The loan was approved because '… the company's output is valuable and cannot be dispensed with'. Repayment was dependent upon the company's plans. It was disclosed that Grahame-White and two other aircraft companies were contemplating floating an Aircraft Trust in 1918 with capital of £2,000,000.

King George V and Queen Mary visited the Grahame-White factory on 4 December 1917, by which time the workforce had increased to more than 2,300.

Post-war Hendon

7

The Armistice of 11 November 1918 was quickly followed by the cancellation of most aircraft contracts. This left Grahame-White with a problem; civil flying was still banned and his factory had no work. Until the Aeroplanes (Experimental Manufacture) Order of 1917 was suspended the company could not even begin building civil aircraft. One of the few companies that still had work was British Aerial Transport Co Ltd (BAT), which had been formed at Willesden by Lord Waring in 1917 and used Hendon for testing its military aircraft prototypes.

Claude Grahame-White sought the return of his aerodrome and the acquisition of the buildings that the Government had constructed on it. As early as 28 November 1917 he had sought compensation from the War Losses Committee for loss of earnings caused by Admiralty activity at Hendon, including the exclusion of the public, and demonstration flying. He had other claims pending for compensation, including one resulting from changes to the flying training contract. He was not alone in this because on 20 December 1921 an action was heard over the treatment of the Ruffy-Baumann School and the cancellation of its training contract. The court found in its favour, but only awarded £250 as opposed to the counter claim of £1,227 awarded to the Crown for goods supplied. In return, however, the Government sought repayment of the loans given to the Grahame-White company. On 27 March 1920 W. A. Robinson of the Air Ministry wrote to the Treasury that the RAF had vacated Hendon and 'desired to come to an early settlement with the Grahame-White company', but until the various claims were settled the Government would not relinquish control.

In 1919 construction began of a clubhouse east of the factory buildings on 'Little Hendon'. In *Flight* for 10 May 1913 there had been mention of the first suggestion of a club and a company, the London Aerodrome Club Ltd, registered to run it. Unfortunately the war had brought development to a close and in 1916 the company was struck off. When construction started a new company, the London Flying Club Ltd, was formed to run it. It was based on clubs Claude Grahame-White had seen in America and featured tennis courts, a ballroom,

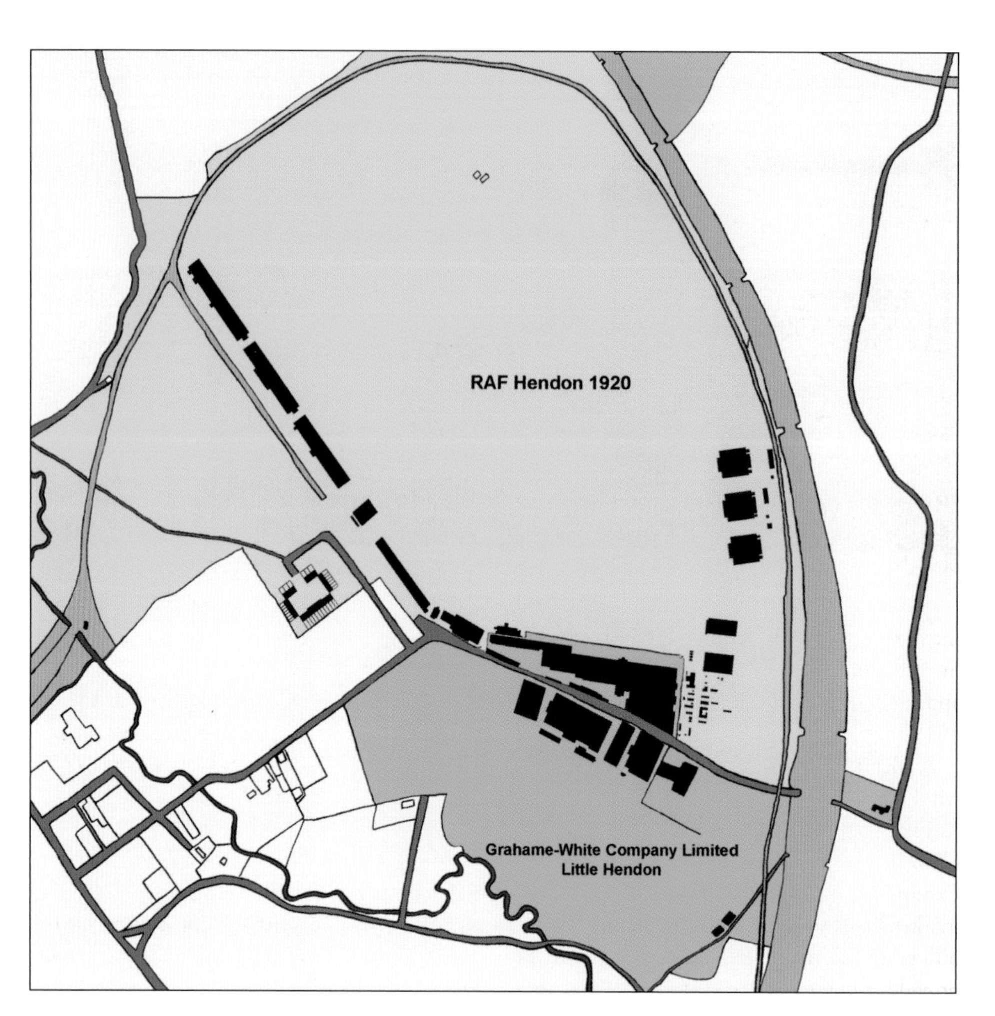

Hendon Aerodrome in 1920.

Grahame-White was able to keep some of his workers busy by making the furniture for the clubhouse.

Most aircraft contracts were cancelled in 1918 except for new types such as the BAT Fk.23 Bantam. Many of them, however, had production contracts reduced or cancelled.

theatre and other attractions. It generated such interest that a short article about the club appeared in the *New York Times* for 4 May 1919.

A furniture business began in 1919 under Mr E. Warr King and produced most of the furniture and fittings in the London Flying Club. It had a stand at the 1920 Ideal Home Exhibition, but did not aim for the domestic market; it was wholesale only, mainly hotels, offices and libraries. However, this could not provide enough work to sustain the Grahame-White factory, so an attempt was made to restart aircraft production.

In 1919 Grahame-White began producing aircraft once more. They were designed by Monsieur Boudot and included the Bantam and Aero-Limousine. Reginald Carr returned, having been a test pilot at Martlesham Heath; he had been chief ground engineer before becoming a pilot for Grahame-White before the war. The aircraft designs attracted little interest, however, and in Jane's *All the World's Aircraft* for 1920 came the announcement that the company 'is no longer constructing aircraft'. In 1922 Carr left the company and founded his own electrical business before moving to Handley Page in 1940.

While the aircraft work dried up Grahame-White turned his attention to car production and opened a car showroom at 12 Regent Street, managed by Leonard Williams.

Reginald Carr returned to Hendon after the war and briefly became a pilot for Grahame-White once more.

Grahame-White is supposed to have bought as many war surplus vehicles on Rolls Royce chassis as possible and rebodied them. Details can be found on-line of some cars that the company rebuilt and were subsequently shipped to Australia and New Zealand. Daimler cars were also fitted with Grahame-White bodywork and some are known to have been exported to India. In addition there is photographic proof of the construction of new bodies on Crossley chassis. The body type is described as the Manchester touring car coachwork as fitted to the 25/30 and the cars had the Crossley bonnet badge. It would appear, unfortunately, that he never put his company's name on his coachwork. At the 1919 Motor Show the Delahaye stand featured a car with patented Grahame-White bodywork. No price was mentioned for a complete car, but the body alone was priced at £650.

Some cars did feature the company logo. The first was the Grahame-White Super Scooter, a British version of the Briggs & Stratton Flyer. By the time of the 1919 Olympia Motor Show the company was offering the Buckboard and 10hp Light Car. The Buckboard may have been based on the Orient Buckboard, which was sold in the UK by Blackfriars Motor & Engineering Works of Stamford. It had a 3hp Baker-Precision engine, and the Light Car had a 10hp 1,100cc Dorman engine. Buckboards were produced at the rate of about 100 per week and enough orders were taken at Olympia for six months' production. In 1920 the Buckboard was replaced by the Wonder Car. It was shown at the 1920 Olympia Motor Show with a 3½hp engine or a 7hp Coventry-Victor engine. The cars were still on sale in 1921, but the company did not attend the Motor Show.

Like many other companies, A. Harper, Sons & Bean Ltd suffered from the cancellation of contracts in November 1918. In order to survive it too turned to motor manufacture, and in January 1919 bought the rights to the Perry 11.9 car; it was redesigned and renamed the Bean 11.9. An order for 2,000 Bean bodies was placed with Grahame-White in 1920, possibly 29 June. Another order was placed for a further 5,000 bodies before completion of the first one. When part payment was requested the second order was cancelled and placed with Handley Page instead. Financial problems may have been the reason for the refusal of payment, because a receiver was appointed in October 1920. Grahame-White was left with 500 unsold bodies, which were altered and sold to Crossley of Manchester, possibly in part payment for the Crossley chassis. In November 1921 new investment allowed A. Harper, Sons & Bean to resume production early the following year. A published directory of car companies does not mention any Crossley car suitable for the former Bean bodies, but it does show a 'Bean body' on a

Manchester-style car bodies being fitted to Crossley 23/30 chassis in the Grahame-White factory.

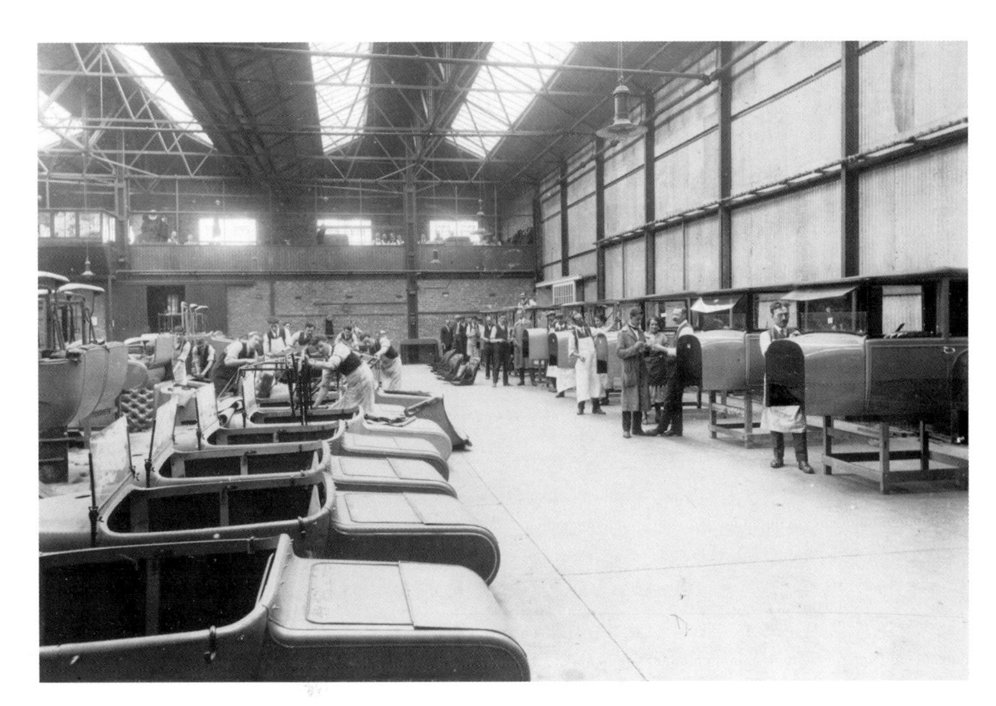

Car bodies under construction for A. Harper, Sons & Bean Ltd.

Fiat 501, a car produced from 1919 until 1926.

The introduction of the Austin Seven in 1922 had a profound effect on the car industry. Several companies had turned to the production of cycle-cars in the post-war years in an attempt to replace contracts lost at the end of the war. These simple vehicles seemed an easy way to introduce motoring to the masses, but the Austin Seven rendered them obsolete. Unfortunately that included the Grahame-White cars, and he was soon under financial pressure once more.

From 1 May 1919 civil flying was permitted once more and almost immediately the first display of that year was held. It saw the arrival of three Blackburn Kangaroo aircraft for Grahame-White to join the four Avro 504s used for pleasure flights. Visitors began to return to Hendon, but the events were not as common as they had been. It was even harder to find pupils to keep the schools working. Clarence Winchester was recruited by Grahame-White as a flying instructor but he resigned with effect from the end of September 1919. Despite this setback the school was one of the few to see a pilot qualify in 1919; the American Barclay Harding Warburton Jnr qualified in September 1919 under instruction from Captain Paul Richard Tankerville-Chamberlayne AFC (later Air Commodore Paul Richard Tankerville James Michael Isidore Camille Chamberlayne). Despite this early success the school struggled, but even as late as 1921 it was still offering flying tuition, £125 for a course leading to an Aviator's Certificate and a further £100 for an Advanced Aviator's Certificate.

The Victory (fourth) Aerial Derby was held on 21 June 1919 and the winner was Capt Gerald William Gathergood of Airco. The progress of the racers was displayed on a special scoreboard located near the pre-war starter's hut and booking office. The French ace Nungesser visited and regretted the absence of anything similar in France. For the 1920 Aerial Derby the judges were in telephone contact with observers at the turning points. The race was organised by the Royal Aero Club, and its members had the use of the London Flying Club, which opened in that year. However, it soon changed its name to the London Country Club because the committee felt the other was too restrictive and did not reflect the changing interests of its members.

Aerofilms was founded by Claude Grahame-White and Francis Lewis Wills and specialised in aerial photography. Wills had been an observer in the RNAS and RAF before being demobbed in 1919. The company was registered on 20 May 1919 with Wills as managing director, Grahame-White as chairman and Herbert William Matthews as third director. Initially it was based in the country club and used aircraft from Airco. When that company closed they used the pilots

A Blackburn Kangaroo of the Grahame-White Co Ltd with the Hucks Starter connected in 1919. Three of these war-surplus aircraft were bought to start an airline service, but they were scrapped after being used for pleasure flights

of its successor, the de Havilland Aeroplane Hire Service, including Alan Cobham. In 1925 Aerofilms was bought by the Aircraft Operating Co Ltd and moved to Aerial House, Colindale.

In 1920 the RAF Care and Maintenance party left Hendon. Almost immediately, however, a new party assembled on the aerodrome, tasked with preparations for the first Royal Air Force Aerial Pageant, held 'by kind permission of the Grahame-White Company' to raise money for service charities. When it was held on 8 July, 40,000 paid on the gate but the crowd was reckoned to be much larger, with all the surrounding vantage points occupied as they had been for pre-war displays. Despite the congestion on surrounding roads it was considered to be one of the best displays ever held. The pageant and aerial derby were the highlights of an otherwise quiet year at Hendon, which also saw Airco and BAT close.

Claude and Ethel were surprised when Georgette Cohen, Ethel's daughter, married broker J. William Souther on 24 February 1921 and gave up the stage in the following month. In 1919 it had been expected that she would

The London Flying Club.

The Victory Aerial Derby of 1919 was won by Captain Gerald William Gathergood flying this Airco DH.4R. Its civil registration of K141 was replaced a month later when the international system still used today was adopted.

perform in *Peter Pan*, but instead returned to her father in the United States. The announcement of their engagement had been expected, but they slipped away and married in secret at Lakeworth, West Palm Beach. Souther died in 1925, and on 15 March 1926 Georgette married William Hamilton Rouse.

In 1921 Claude and Ethel sailed from Southampton to New York, where he sought as many new products as possible. His company became the sole British agents of the Turn-Auto, an American device that allowed cars to be rotated for ease of access to the chassis and was used in the Rolls Royce factory in Springfield, Massachusetts. He also became agent for the Hickman Sea Sled motor boat, developed by William Albert Hickman, a Canadian. It first appeared in 1913 and polarised opinion, not helped by Hickman's attitude, which irritated his contemporaries and won him few friends. Claude made a number of trips with George Leary Jnr in *Orlo II*; Leary, from New York, had won a number of trophies with the Sea Sled.

November 1922 saw a worrying headline in *Flight*. The acquisition of some land for the

Bristol Fighters over the Grahame-White factory. The first RAF Aerial Pageant was held in 1920 and raised money for the RAF Memorial Fund.

extension of the London Underground had led to a misunderstanding. Journalists assumed that the whole of Hendon aerodrome had been bought and was to be turned into a housing estate. Just one week after the announcement *Flight* was pleased to correct the misunderstanding on its front page. Indeed, it mentioned the additional land that Grahame-White had purchased from the estate of the late Sir John Blundell Maple Bt. This area was the land between the aerodrome and Edgware Road north of Colindale Avenue, including all the land over which the railway spur ran. *Flight* preferred Hendon over Waddon and felt that the former should be London's airport, the new land being used for a 'sports aerodrome'. In 1923 the extension of the Underground from Golders Green to Edgware cut across 'Little Hendon', isolating the part adjacent to Colindeep Lane. The remainder was unusable as an aerodrome, so was converted into a golf course.

In 1922 Australian Richard Tilden Smith offered to buy Grahame-White's company and property. The initial offer was absurdly low, but gradually increased during months of negotiation. Tilden Smith sought the entire estate, but Grahame-White wished to retain part. Various draft agreements were prepared and Claude sought the advice of his uncle, Lord Barnby of Blyth Hall, Nottinghamshire, who also knew Tilden Smith. While negotiations were taking place, Grahame-White was still hopeful that Hendon could become the new London Airport instead of Croydon, and knew that the extension of the Underground from Golders Green to Edgware would increase the value of the property. Correspondence held by the RAF Museum ceases in January 1923, implying that negotiations had failed.

Financial pressure resulted in the Aerial Derby being taken away from Hendon in 1922 and moved to Croydon. Grahame-White sought too high a fee from the Royal Aero Club for Hendon to be used as either the start/finish or as a turning point. *Flight* bemoaned the lack of access to Croydon by public transport and the risk of airport users hampering the racers, but recognised the need of the Royal Aero Club to be financially prudent in the current economic climate. By now, however, interest in the Aerial Derby was waning, and it was not the attraction it had been. The RAF Pageants still managed to attract the public and keep Hendon in the public eye.

By December 1922 Grahame-White had orders for five Sea Sleds, four of which had Studebaker engines. Two of these motor boats were built by S. E. Saunders at Cowes. Another, of 28 feet and named *Charabon*, was built in 1923 by James Taylor & Bates at Chertsey and fitted with a Rolls Royce Eagle aero engine. No one seems to have noticed that the Grahame White company was also making prams. The 'pram' concerned, however, was the proper name for an 8ft 6in dinghy for use on a yacht, carried on davits, and praised for its ease of use.

Grahame-White also kept himself busy. In July 1923 he competed in the British Motor Boat Club races on the Thames near Chelsea. He used the 32-foot boat *Ispano*, fitted with a 40hp Hispano-Suiza engine, and finished second in two races. He was also a member of the Royal Motor Yacht Club. In December 1923 he stated that he was in negotiation for the sale of 20 acres for a wealthy continental company to build a factory for the production of goods upon which a protective tariff had been imposed. With his factory idle and school closed he needed other sources of revenue, and the simplest way was to lease buildings to other companies.

The company's income at the time was low, most of it coming from rents, with additional income from the sale of furniture and the occasional car. At the time of the appointment of the receiver the largest expense for the company was its wages bill. H. E. Hutchins received £46 per month, Bernard Isaac (who had returned to the company) £40 per month, and H. P. G. Brackenridge £33 6s 8d. In addition, the weekly wage bill at Hendon was £80 14s 6d and falling, while the weekly wage bill at 12 Regent Street fell during March 1924 from £17 5s to £4, suggesting a dramatic reduction in the number of staff.

As early as 1920 Grahame-White had leased some of the former Acceptance Park hangars to W. C. Gaunt, importer of Packard cars and lorries. With the closure of the Grahame-White factory, better facilities were available and Gaunt moved into other buildings, with Leonard Williams joining him as manager. Another early resident was Savage Skywriters, founded by Major John Clifford (Jack) Savage using obsolete SE.5a fighters. He had worked

The failure of aircraft and vehicle production meant that only furniture production remained, insufficient to keep the factory busy.

for Grahame-White, then B. C. Hucks, before joining the Army during the First World War. He was demobbed with the rank of Major and began working for Handley Page. His earlier work with Hucks led him to seek a method of generating smoke; he patented a number of methods and had developed a successful system by 1922. The method used was very similar to that used today by the Red Arrows aerobatic team and was used to write advertising slogans in the sky.

When Grahame-White closed his factory most of it was leased to two car companies. Tylor Engineering Co Ltd was formed on 9 May 1922 to acquire the engine manufacturing business of J. Tylor & Sons Ltd; the company leased the Hendon factory for £5,749 18s 10d in 1923. On 10 October of that year, however, Sir Albert William Wyon was appointed as receiver on behalf of Lloyds Bank. At the time he was appointed the company received £507 for sub-letting part of the factory, but this had

One of the Savage Skywriters' aircraft flying low, now owned by the Shuttleworth Collection.

Any source of income was welcome, including rent for grazing animals or haymaking.

dropped to £10 10s by 10 April 1924, and it surrendered the lease soon after.

Like many other munitions companies, Sir Wm Angus Sanderson & Co Ltd of Newcastle planned to enter car production, but on 27 January 1921 Joseph Sankey & Sons Ltd petitioned for its winding-up. A new company, Angus Sanderson Ltd, was incorporated on 13 January 1922, purchased the remaining assets of its predecessor and planned to restart production in Hendon, using engines made by Tylor Engineering. The company was soon caught up in the same financial problems as Tylor, however, and on 18 October 1923 Wyon was appointed as receiver of Angus Sanderson too. His work with the company ceased on 26 October 1923, and 26 November it confirmed to the Registrar of Joint Stock Companies that it was still active as motor car manufacturers with its works at London Aerodrome. On 21 December 1923, however, James Tait was appointed as receiver. By February 1924 the Grahame-White company secretary had written off any chance of recovering money owing when corresponding with the receiver. There is nothing to indicate that any cars were made in the Grahame-White factory.

The problem for both Tylor and Angus Sanderson was their involvement with Colonel Richard Durand Temple of the Industrial Guarantee Corporation Ltd. He had a plan whereby car manufacture could be insured by

By 1923 others occupied Claude Grahame-White's former office.

Lloyd's of London, and he convinced Stanley Price Knowles Harrison of Lloyd's to join him. Unfortunately the scheme was based on false information and open to fraud, so when Harrison tried to recover his losses from Lloyd's he found he was powerless to do so and was forced, like his companies, into bankruptcy.

1923 was a strange year for Hendon. It was the calm before the storm, but that did not stop things happening. Hendon Council acquired Sunny Fields, and in 1923 granted the company control of it so that any admission charge would benefit charity; profits from the Pageants were given to the Royal Air Force Benevolent Fund. No more screens were needed like those erected in 1914. Some post-war users of Hendon were not quite as respectable and went to jail because of their activities. Fraser's Flying School opened at Kingsbury in 1922 under the supervision of Alexander Fraser. It was thought it would move to Hendon but does not appear to have done so, and Fraser's fraudulent claims resulted in him being sentenced to 18 months imprisonment with hard labour on 6 October 1923.

On the morning of 21 February 1923 a fire started in one of the former Acceptance Park sheds used by Messrs Cole & Sons (1923) Ltd for varnishing the wood of car bodies. The fire spread quickly to three other sheds; one was used by Cole for the storage of materials and eleven new cars, the other two were used by Dugdale for the storage of tractors and other agricultural machinery. Engines from seven fire stations were needed to control the fire and stop it spreading. An empty shed helped prevent the fire from spreading to sheds storing forty to fifty buses of the Cambrian Motor-Coach Company, and a narrow passage divided the destroyed sheds from those used by General Motors for the storage of cars.

General Motors started importing vehicles to Britain in 1923. Import duty made their vehicles expensive to buy, so it was decided to start making them in England, and production of Chevrolet commercial vehicles started in part of the former Airco works on the Edgware Road. In 1925 General Motors bought Vauxhall and production eventually moved to Luton, and the commercial vehicles were renamed Bedford. It is unclear whether the vehicles mentioned were imported or built locally.

In order to promote civil aviation the Air Ministry held light aeroplane trials at Lympne in order to find aircraft to equip new flying schools. On 27 October 1923 the entrants were exhibited at Hendon at the last major event of that year. The Aerial Derby was moved once more to Croydon, but the RAF Aerial Pageant was held in June, and in July an event new to Hendon, the Kings Cup Air Race, was held, similar in format to the pre-war round-Britain races. The first Kings Cup Air Race in 1922 had started at Croydon.

In 1923 the Treasury was ready to come to a settlement with Grahame-White. The Air Ministry, however, had reconsidered its

The Gloster Grebe in the New Types Park at the 1923 Pageant, the first year to feature this attraction.

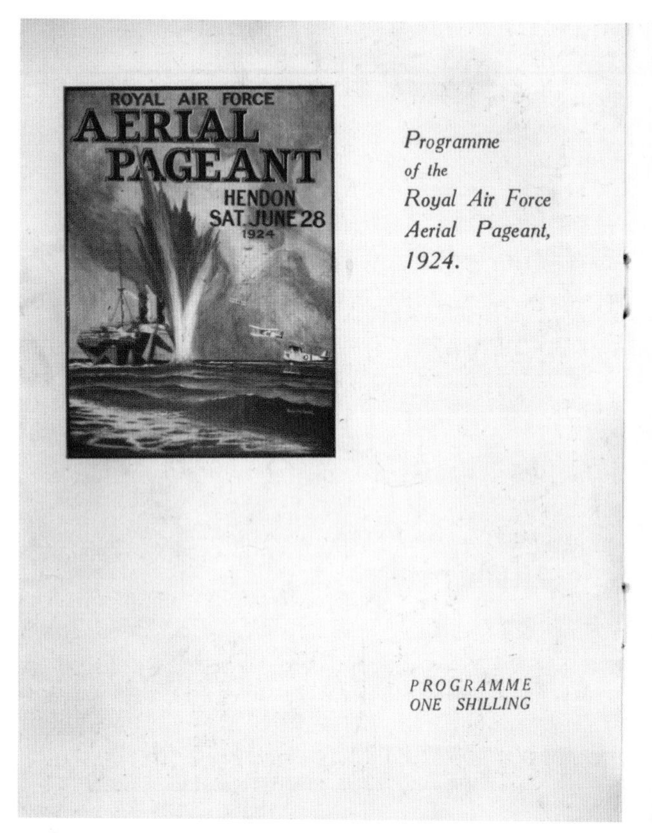

Programme
of the
Royal Air Force
Aerial Pageant,
1924.

PROGRAMME
ONE SHILLING

Royal Air Force Aerial Pageant

(Held in connection with the Royal Air Force Annual Training)
In Aid of
THE ROYAL AIR FORCE MEMORIAL FUND
Held at
THE AERODROME, HENDON
at 3 p.m. SATURDAY, JUNE 28th, 1924.
Gates open 10.30 a.m. Preliminary Flying Heats from 11 a.m. R.A.F. Bands will play from 1 p.m.

PROGRAMME

PATRON - - - - - H.R.H. THE DUKE OF YORK, K.G., G.C.V.O.

PRESIDENT.
Air Chief Marshal Sir H. M. TRENCHARD, Bart., G.C.B., D.S.O., A.D.C.

VICE-PRESIDENTS.
Air Vice-Marshal Sir V. VYVYAN, K.C.B., D.S.O.
Air Commodore C. L. LAMBE, C.B., C.M.G., D.S.O.
Air Commodore C. A. H. LONGCROFT, C.B., C.M.G., D.S.O., A.F.C.
Air Commodore E. A. D. MASTERMAN, C.B., C.M.G., C.B.E., A.F.C.
Air Commodore L. E. O. CHARLTON, C.B., C.M.G., D.S.O.
Air Commodore C. R. SAMSON, C.M.G., D.S.O., A.F.C.
Air Commodore A. E. BORTON, C.B., C.M.G., D.S.O., A.F.C.
Group Captain H. V. WELLS, C.B.E.
Commander H. E. PERRIN, R.N.V.R.

CHAIRMAN.
Air Commodore T. I. WEBB-BOWEN, C.B., C.M.G.

SECRETARY.
Squadron Leader F. L. ROBINSON, D.S.O., M.C., A.F.C.

AERODROME ORGANISER.
Flight-Lieutenant R. HALLEY, D.F.C., A.F.C.

FLYING COMMITTEE.
Air Commodore H. C. T. DOWDING, C.M.G.
Squadron Leader R. M. HILL, M.C., A.F.C.
Flight-Lieutenant C. R. CARR, D.F.C.

GENERAL COMMITTEE.
Squadron Leader C. C. DARLEY, A.M.
C. P. ROBERTSON, Esq.
Squadron Leader J. H. PEEK, M.D.
Squadron Leader R. S. OVERTON.
Flight-Lieutenant W. C. CLARK.
Flight-Lieutenant D. Gilley, D.F.C.
Flight-Lieutenant D. E. WARD.
Flying Officer R. S. MARTIN.

JUDGE.
Group Captain I. M. BONHAM CARTER, C.B., O.B.E.

The Committee wish to express their thanks for the valuable co-operation and assistance received from so many sources, and in particular to convey their acknowledgments to :—

1. The Stoll Film Co., who voluntarily offered, designed, erected and painted the ships used in Event No. 9.
2. Viscount Rothermere and Colonel Ivo Fraser, for generously providing free advertising facilities.
3. The press generally, for the wide and sympathetic publicity which has been given to the Pageant.
4. The London Electric Railway Companies, London United Tramways, Ltd., London General Omnibus Co., and the London Midland & Scottish Railway Co., for the free display of Pageant posters, and for affording special transport facilities.
5. Messrs. J. Waddington & Co., Ltd., and Mr. A. B. Ward, of the British Periodicals, Ltd., who undertook the general printing arrangements for the Pageant.
6. The Metropolitan Police Force, for their co-operation in organising the traffic scheme.
7. The London Country Club, Hendon, N.W., for the hospitality given to members of the Royal Air Force whilst at Hendon.
8. All Libraries and Booking Agencies for special facilities provided for the advance sale of tickets.
9. The Shepherd's Bush Exhibition, Ltd., Wood Lane, who kindly supplied free 30 additional turnstiles for the entrances.
10. Major J. C. Savage, for kindly putting at the disposal of the Pageant Committee part of a shed on the aerodrome.
11. Queen Alexandra's Hospital, Millbank, The Medical Officer of Health, Hendon Urban District Council, St. John's Ambulance Brigade (No. 56, Cricklewood) Division, and Colindale Hospital, Hendon, for kindly co-operating in the medical a rangements.
12. The Hendon Urban District Council, The Metropolitan Water Board, The Middlesex Gas Co., for their great efforts to clear the various approaches which have recently been under construction.
13. Messrs. Rolls-Royce and D. Napier & Son, Ltd., for their expert assistance on the engines used in machines for the Pageant.
14. Messrs. Cardinal and Harford, Ltd., for the loan of a quantity of rare Persian Rugs for the Royal Enclosure.
15. Messrs. Heelas & Sons, Reading, for carrying out the decorating of the Royal Enclosure at a reduced price.

REFRESHMENTS BY
MR. W. POLLARD, 41, Chobham Road, Woking.
Lunch and Tea will be served as follows :—
10/- Enclosure. — In the Hotel and also in the Restaurant close to the Main Entrance : Lunch, 5s. ; Tea, 2s. per head.
5/- Enclosure. — Buffet lunch and teas in the Restaurant.
Both 2/- Enclosures. — Buffet Refreshments.
Alcoholic Refreshments can be obtained in all the Enclosures at separate Bars.

NOTICE.—The Committee have not authorised any collections to be made within the precincts of the aerodrome.

The programme for the 1924 Pageant, the last to bear the name.

position regarding the retention of Hendon and now wished to use it for the planned Auxiliary Air Force. Negotiations between Grahame-White and the Air Ministry were announced in the aviation press in April 1924, but by then the Government had acted and appointed a receiver. The company owed the British Government more than £330,000, with debtors including the Admiralty, the Treasury and the former Ministry of Munitions. Repayment was sought and, after taking legal advice, the Government appointed Mr David Livingstone Honeyman as receiver in a letter dated 29 February 1924, the day he took possession of the factory, and confirmed in court on 6 June. Grahame-White then started a counter-claim in court against the Treasury Solicitor on 29 May, and from this moment negotiations proceeded slowly.

In 1924 a new flying club was planned for Hendon, the London Flying Society, by people well known at Hendon, Cecil Pashley and Clarence Winchester. It would have been based at Hendon and had already acquired a 80hp Avro. Unfortunately it appears to have been a victim of the appointment of a receiver due to the loss of Hendon. The Royal Aero Club was also unable to use Hendon for events. The only one that did take place was the annual RAF Aerial Pageant, the last that would bear that name.

The intention of the Treasury was that the receiver would clear the site, and surplus land could be sold with the minimum being retained for use by the RAF. Unfortunately events did not work out as planned. General Motors, Angus Sanderson, Tylor Engineering and Delco Lighting were all given notice to quit, and had done so by the end of 1924. Many of the others, however, remained on site, including W. C. Gaunt, Savage Skywriters, Alfred Dugdale Ltd and Cambrian Coaching & Goods Transport Co Ltd.

It was hard for the receiver to find a buyer, so he decided to lease the buildings. A new tenant was the Franco British Electrical Co Ltd, founded by Henry Bey. Like many companies at the time, it took its name from the Franco-British Exhibition, organised by Imre Kiralfy at his exhibition site at White City in 1908, and was responsible for many of the illuminated advertisements in Piccadilly Circus. In 1910 Western Electric Ltd was incorporated as the UK arm of its American parent, and in 1922 it

Opposite page:
Franco British Signs was responsible for many of the illuminations in Piccadilly Circus.

STC used Hendon, among other things, for the development of aircraft radio. Equipment was manufactured for both aircraft and ground stations.

bought the Tylor factory in New Southgate. Now that the company needed more space it planned to buy part of Hendon. In 1925, however, the company was bought by ITT and renamed the Standard Telephones & Cables Co Ltd (STC). The new owners preferred to lease the buildings, so an agreement was reached. It was not long before the company was using almost all of the factory buildings, with the exception of the one occupied by Franco British.

Events on the airfield in 1925 were few. The RAF Aerial Pageant was replaced by the RAF Display in June, but it was as popular as ever. Later there was an evaluation by Sqn Ldr Haig and Flg Off Merer concerning the suitability of Hendon for test flights of experimental aircraft. RAE Report K2036 dated 20 August 1925 concluded that test flights could be conducted in a NW-NNW direction; this was probably no surprise for Grahame-White and the other companies that had used Hendon for first flights.

Negotiations between the receiver, Grahame-White and the various Government departments continued through 1925. Unfortunately, while this was going on his company, together with others, was sued by Robert Esnault-Pelterie. He claimed that he had invented the 'joystick' and that the companies owed him a royalty on every one used in British aircraft in France during the First World War. In April 1925 the French court fortunately ruled in favour of the companies, pointing out that the British Government had specified that the aircraft should be fitted with joysticks and that he should have sued the British Government instead. Although the claim against Grahame-White would not have been as large as that faced by many of the other companies, it would have been yet another bill his company could ill afford.

It must have been difficult for Claude to concentrate on the negotiations because of his mother's health. On 7 December 1925 he attended her funeral at Roffey Church, Horsham. On the following day the agreement was signed defining the settlement between Grahame-White, his company, the Treasury Solicitor, Mr Honeyman and the Air Council. Further negotiations were needed with the leaseholders, but in 1926 the transfer of ownership was complete. Grahame-White's interest in the London Country Club was also purchased as part of the 1925 settlement, and on 31 December of that year the club was given notice to quit. Closure finally took place on 30 January 1926 after 6½ years of operation.

8 RAF Hendon

When the Government took control there were still a number of tenants at Hendon, including STC, Franco British and Savage Skywriters. Other properties included Aeroville, Canberra House, which was occupied by H. E. Hutchins, Grahame-White's company secretary, Featherstone Farm, Hillfoot House and Finchley Football Club. Their tenancies were transferred to the Air Ministry Directorate of Lands with effect from 23 September 1926, and on 10 January 1928 Sir Samuel Hoare was able to write to the receiver to congratulate him on presentation of the final accounts and the completion of the acquisition of Hendon.

The RAF Display for 1926 was held on 3 July. Now that Hendon's future was secure, the Royal Aero Club returned with the Kings Cup Air Race a week later. The event was a handicap race held over two days, 9-10 July. Once the events were over, work began to improve the drainage of the airfield. In addition the railway spur was lifted and its ballast used to fill an area of boggy ground near the former Aerodrome Hotel.

The 1926 display marked the end of an era. Volunteers arrived at Hendon from various RAF units around the country and lived in a camp on the aerodrome for about three months, preparing it for the display by building the set pieces and erecting the fences for the various enclosures. Although the camp had an officer in charge and an adjutant, discipline was relaxed and everyone looked forward to it, despite the basic facilities. Once Hendon had its own permanent staff the volunteers preparing for the 1927 Display found the atmosphere changed and discipline far stricter.

The first permanent residents of RAF Hendon came with the formation of the Auxiliary Air Force (AuxAF). In 1924 it was planned that Hendon would accommodate two squadrons, with camp buildings being needed from 15 January 1925 and the aerodrome and two hangars being needed from 31 January. The final requirement was the Aerodrome Hotel, which would be used as the officers' mess and quarters from 2 March 1925. The delay in the purchase of Hendon meant that 600 and 601 Squadrons formed at Northolt in 1925 as bomber squadrons. It was also planned to form the first Special Reserve Squadron, 500 Squadron, at Hendon, but when it finally formed in 1931 it did so at RAF Manston.

In January 1927 600 and 601 Squadrons moved to Hendon and became the station's first residents. Two hangars housed the aircraft while the one between them was converted into offices and workshops. RAF Station Hendon, however, did not come into existence until 1 December 1927, when Wing Commander Crosbie, Superintendent of Reserve, moved in from Northolt and combined his duties with those of Station Commander. He was assisted as Superintendent by two squadron leaders and a flight lieutenant. The station officers consisted of two flight lieutenants: adjutant and accounting officer. A medical officer was not appointed until March 1928.

The activities at RAF Hendon resembled its pre-war days. Flying by the AuxAF squadrons was undertaken on Thursday evenings during the summer and at weekends. Night flying was also practiced on Thursday evenings in the summer. Training was given at the squadrons' headquarters every Monday evening and on Thursday evenings during the winter. The squadrons' headquarters closed on Tuesdays and Wednesdays in order to give the permanent staff time off instead of at weekends.

Once the RAF Display was over the auxiliary squadrons could leave Hendon for their annual summer camps, usually in early August. 601 Squadron, for example, often went to Lympne, close to the home of its Commanding Officer, Sir Philip Sassoon. These camps provided two weeks of intense training for the whole squadron of about 150 officers and men.

The former RNAS buildings had been the only quarters on the station for several years, but were in a poor condition. A period of construction began with the building of a new station headquarters and accommodation for officers and men. The remaining pre-war sheds were demolished and married quarters built alongside a new road, which in 1933 was named Booth Road (the original Booth Road was renamed Sheaveshill Avenue). This work was followed soon after by the construction of technical buildings in the south-east corner,

This camp from 1928 is typical of the conditions RAF personnel faced while preparing for the annual RAF Display.

When the RAF took control of Hendon the former hotel was the only office accommodation available to it.

replacing the offices and workshops in one of the hangars. Once construction was completed it enabled the formation of a third AuxAF squadron, 604 Squadron, which formed at Hendon on 17 March 1930. The main function of the Superintendent of Reserve was the control of RAF reserve forces, including the Auxiliary Air Force. One could say, therefore, that Hendon was not just home to three squadrons, but also The Home of the Auxiliary Air Force.

Even at the end of the 1920s the Treasury still wanted to maximise its return from the Hendon estate and recover the money it had paid for the site. At the end of the First World War the Government had promised 'homes fit for heroes', and passed a new Housing & Town Planning Act in 1919, making it a statutory duty for local authorities to provide housing for the working class. Space in London was limited, so the London County Council was forced to look further afield.

The decision to build the Watling Estate, close to the new Burnt Oak tube station, was taken in 1924, and construction began in 1927. The Treasury took the opportunity to sell part of the Grahame-White estate outside the aerodrome boundary for incorporation in the estate, and in this they were assisted by the acquisition of two further pieces of land from Grahame-White in 1929. Their acquisition enabled management of a stream and subsequent use of the land for housing, improving its value. Fortunately for residents in the area, large parts were used as parks and playing fields, assisting the RAF by avoiding obstructions too close to the aerodrome. The Featherstone Farm was another part that was sold and, while Franco British bought its part of the former Grahame-White factory, most of it was still leased to STC. Savage Skywriters still managed to retain a presence on the airfield despite the United States having become the company's main market. Despite this apparent move overseas some operations seem to continue at Hendon until about 1931. In 1932 Bert Hinkler used one of Savage's sheds while working on his Puss Moth Karohi, but they appear to have been underused in the 1930s.

In 1932 Alfred Dugdale Ltd went into voluntary liquidation and relinquished its

A recruitment poster for 600 Squadron AuxAF designed by W. E. Johns, creator of 'Biggles'.

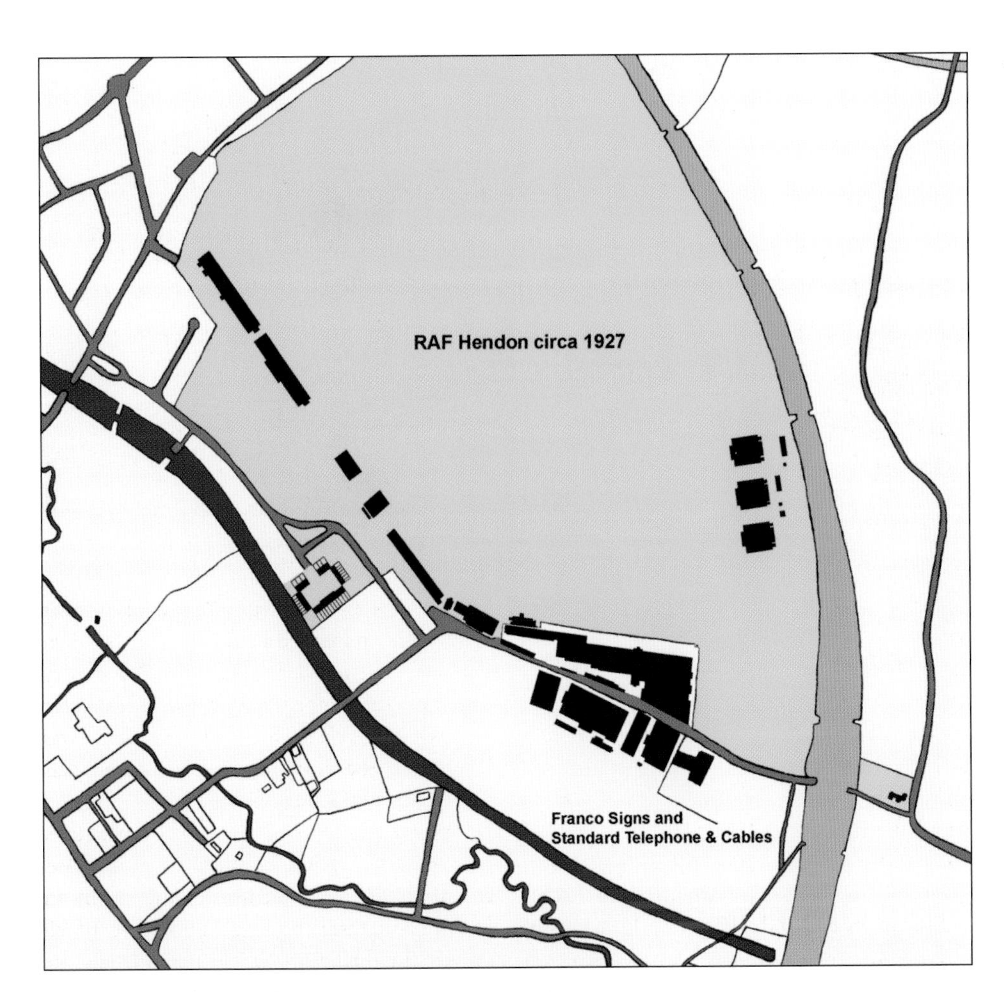

RAF Hendon circa 1927.

The Prince of Wales in his role as Honorary Air Commodore-in-Chief Auxiliary Air Force inspects men of 601 Squadron at their headquarters in Notting Hill, circa 1932.

Some of the aircraft of 24 Squadron, including the Prince of Wales's Vickers Viastra in the background and the Avro Cadet of the Secretary of State for Air in the foreground.

sheds at Hendon. The availability of this accommodation finally allowed 24 Squadron to move there on 10 July 1933, something planned as early as 1924. When it did so it absorbed the Home Communications Flight, which had been at Hendon since 16 April 1928. RAF officers had also been permitted to keep their own aircraft at Hendon, and from 1930 so too did the Prince of Wales. 24 Squadron had provided aircraft for official duties, but his aircraft were used for private visits. Unlike the other officers, however, 24 Squadron provided support for them.

Two separate camps had formed at Hendon comprising separate barracks, institutes, sergeants' messes, etc: East Camp mainly housed the Auxiliary Air Force, while West Camp housed the regular RAF personnel, although many RAF officers were quartered in Canberra House and Hillfoot House. The RAF and AuxAF officers each had their own mess, while the former Aerodrome Hotel was the Auxiliary Air Force officers' mess.

The lease to STC had been renewed in 1929 but was due to expire in 1934. By then, however, the company had redeveloped the site at New Southgate and was able to consolidate there, releasing the factory back to the Air Ministry. The Treasury remained determined to sell as much of Hendon as possible, but was thwarted once more. Negotiations between the Air Ministry and the Metropolitan Police resulted in 'Little Hendon' being taken over by the latter, with the former country club opening on 31 May 1934 as the Met's new training college. The Treasury informed the Air Ministry that if no use could be found for the remaining factory buildings they would be sold. The Air Ministry did identify a use for them, however, and took the opportunity to centralise the storage of reserve vehicles with 1 MT Storage Depot forming at Hendon on 1 August 1935. Initially it was to hold 100 vehicles, rising to 350 by 1 January 1936. It did not stay for long, leaving on 27 November, but its place was taken by 'A' Equipment Depot, which formed on 2 February 1937 for clothing stores.

The expansion of the RAF in the 1930s resulted in a major change to its structure, which in turn had its impact on Hendon. RAF expansion plans meant that pilot training had to increase. Initial pilot training was contracted out to civilian flying schools around the country, while the RAF flying training schools concentrated on the more advanced aspects. It was an arrangement very similar to that seen at Hendon during the First World War.

On 1 July 1935 the roles of Superintendent of Reserve and Station Commander were separated, and Wing Commander Charles Cleaver Miles took command of Hendon the next day. The post of Superintendent of Reserve was renamed Superintendent of the Reserve and Inspector of Civil Flying Training Schools on 1 June 1936, and placed under the control of Training Command, which had formed on 1 May. On 1 December 1937 Headquarters Superintendent of the Reserve and Inspector of Civil Flying Training Schools became Headquarters 26 (Training) Group and administered the Elementary Flying Training Schools. In addition, anyone in the country could write to the AOC 26 (Training) Group for details of how to join the RAF Volunteer Reserve, which had been established in January 1937.

All of the Auxiliary and Special Reserve squadrons had been under the operational control of 1 (Air Defence) Group as part of the Air Defence of Great Britain (ADGB). On 1 May 1936 it was renumbered 6 (Auxiliary) Group, and control was transferred to Bomber Command when it formed on 14 July 1936. On 1 December 1936, however, the squadrons and station were transferred to 11 (Fighter) Group Fighter Command. Hendon's auxiliary squadrons had been redesignated as fighter squadrons as early as 1934, but it took some time to re-equip them. 604 Squadron converted to Hawker Harts in July 1934 and 600 and 601 Squadrons in August, until they were finally replaced with Hawker Demon fighters in 1937.

In 1936 the Prince of Wales became King Edward VIII on the death of his father, King George V. The informal arrangement regarding his aircraft changed with the appointment of Wing Commander Edward Hedley Fielden as Captain of the King's Flight. In December the King abdicated and his place was taken by the

Inside the station stores.

Duke of York as King George VI. In 1937 Edward's private aircraft were replaced by service aircraft and the 1937 RAF Display, the last to be held, was also King George's Coronation Display. The AOC of Fighter Command, Air Chief Marshal Hugh Caswall Tremenheere Dowding, was responsible for the organisation of the display, for which he was appointed Knight Grand Cross of the Royal Victorian Order.

The RAF Display was a victim of its own success. Permanent grandstands had been built, enclosures laid out and public toilets constructed, but Hendon's personnel could not

Refuelling a Demon fighter of 601 Squadron at Hendon.

The 1937 Coronation Display, showing the crowds the events attracted.

cope. Additional men were posted in for the occasion and housed in tented camps; even these were provided with permanent facilities. From 1923 onwards the RAF Pageant had included a New Types Park, but in 1932 the Society of British Aircraft Constructors (SBAC) sought to take advantage of Hendon's facilities and approached the Air Ministry for permission to use it for an invitation-only aero show and flying display. The SBAC display was held on the Monday following the RAF Display, which gave little time to prepare the airfield. It soon became apparent that it was too difficult to accommodate both, and in 1936 the SBAC moved to Hatfield.

On 31 December 1937 the number of officers in the station headquarters was nine, including retired officers and a civilian education officer. Unfortunately there is no information as to the number of men under their command. The number of vehicles available was surprisingly small. There were two staff cars, three 5cwt vans, three 30cwt tenders and two 3-ton tenders. The airfield had one

The SBAC Display at Hendon in 1933.

The auxiliary squadrons at Hendon re-equipped with Blenheim If fighters in time for the 1939 Empire Air Day display.

ambulance, one fire tender, one mobile crane and one tractor. In addition there were two two-wheel flat trailers, one floodlight trailer and a photographic trailer. There would have been other vehicles allocated to the squadrons, but most of the work still had to be done by hand.

The remaining displays at Hendon were much smaller events that did not require men from other stations. On 24 May the country celebrated Empire Day, and from 1934 Empire Day displays were held at RAF stations across the country, attracting potential recruits and promoting the work of the RAF. The first held at Hendon was in 1935, with a second in 1936. The preparations for the Coronation Display prevented one being held in 1937, but they were held in 1938 and 1939. Many mourned the loss of the RAF Displays, but the mood of the country was changing and the RAF had more important matters to consider.

The Munich Crisis of 1938 had the country talking about *when* the war would start, not *if* a war would start. Preparations had begun with the expansion of the RAF in the 1930s, but there was more still to be done. At the end of September 1938 a major exercise was held and the three auxiliary squadrons were dispersed to their war stations; all returned to Hendon on 3 October. In January 1939 they began to re-equip for the final time before the outbreak of war, Blenheims replacing the Demons. The Blenheim If twin-engined fighter was far more complicated, so a Blenheim Conversion Flight was formed to assist.

By January 1939 it was recognised that accommodation at Hendon was a major problem. The planned establishment for corporals and airmen at the station was 418 regulars and 627 auxiliaries. Unfortunately this meant that further accommodation was required for 393 regulars and 418 auxiliaries. The flats in Aeroville were rented privately, and it was realised that as they were vacated they could be used as married quarters, and the top floor could be converted for use as airmen's quarters. Extra was still needed, however, so construction of a hutted camp began in East Camp. Accommodation for aircraft was also a problem, with the First World War sheds considered an obstruction and nearing the end of their useful life. Construction of three Bellman hangars was started on the eastern side of the airfield beside the railway line to provide replacement accommodation.

On 1 February 1939 a further period of reorganisation began. Headquarters Reserve Command was formed at Hendon on 1 February 1939. On the same day Headquarters 26 Group was renumbered 50 Group and placed under its control. On 1 March 1939 HQ Reserve Command moved to White Waltham and on 3 April 1939 HQ 50 Group moved to 11 Tavistock Place, London. In addition 'A' Equipment Depot moved to Wembley on 28 June 1939, by which time it had been renamed A (Temporary) Maintenance Unit. This left just 24, 600, 601 and 604 Squadrons, the Blenheim Conversion Flight and the King's Flight at Hendon.

9 The Second World War

By August 1939 war was inevitable. On 24 August the Auxiliary Air Force was embodied, bringing it under full RAF control, although this was not announced until 1 September when the General Mobilisation was called. On 25 August 600 Squadron moved to its war station at Northolt, followed on 2 September by 601 Squadron to Biggin Hill and 604 Squadron to North Weald. At 11.15am on Sunday 3 September 1939 war was declared against Germany. The King's Flight moved to Benson on 15 September 1939, leaving 24 Squadron and the Blenheim Conversion Flight at Hendon. The only other unit present was the Air Ministry Mechanical Transport Section, which provided transport from Hendon to headquarters in London. On 17 September, however, the Air Despatch Letter Service started, the service being provided by 24 Squadron; Hendon was becoming an important hub for RAF and Air Ministry communications.

A Bristol Blenheim If of 601 Squadron, just before the declaration of war.

As war had drawn closer many of the structures associated with the Hendon displays had been demolished. On 4 September 24 Squadron had moved into the hangar vacated by the departure of 601 Squadron and the remaining rows of First World One sheds were demolished. By then the three Bellman hangars had been completed, but the accommodation huts were not completed until 9 November.

The reduction in resident units enabled Hendon to be used for the formation of a number of new units. RAF elements were sent to France as part of the British Expeditionary Force and Allied Air Striking Force. Between September and October 1939 officers and men of the Special Fighter Cadre Unit and 62

Plans for new accommodation for the King's Flight were cancelled at the outbreak of war.

Fighter Servicing Wing were posted to Hendon before travelling to France. On 31 October 67 Fighter Servicing Wing AASF, and on 1 November 52 Wing BEF, began to form at Hendon before they too proceeded to France later that month. Units from Hendon were not just sent to France; on 1 March 1940 3 W/T Transportable Unit was formed before being sent to Narvik in Norway on 6 May. The Autogiro Unit was the one unit formed at this time that did not go overseas. It was formed circa 1 December 1939 with two civilian-operated Cierva C.30 autogiros for radar calibration, and moved to Martlesham Heath on 20 May 1940.

When German forces occupied the Low Countries in 1940 Hendon acquired new tenants. The Dutch Royal Family moved their aircraft to Hendon and an Allied Flight was formed in 24 Squadron to provide transport for the Belgian & Dutch Army Headquarters.

When France fell 81 Squadron, a communications squadron with Tiger Moths, reassembled at Hendon on 2 May before disbanding on 15 June. 2 Air Formation Signals Unit of the British Army also moved to Hendon after the fall of France, and stayed until it moved to North Africa in 1942. 13 Air Formation Signals Unit later formed at Hendon and took part in the liberation of Europe; these units provided communications between British Army and RAF headquarters.

While the various units had been formed for service overseas, Hendon remained an active Fighter Command Station. The Blenheim Conversion Flight trained thirty-four pilots with only one failure before it was disbanded on 16 January 1940. Another Blenheim unit,

A Miles Magister at Hendon. Aircraft such as these were often attached to RAF units for communications and training.

The officers and men of 248 Squadron, December 1939.

248 Squadron, was formed on 30 October 1939 before moving to Northolt on 26 February 1940. It was intended to be a night-fighter squadron, but the aircraft were found to be unsuitable and it eventually joined Coastal Command. In its place 257 Squadron was formed on 17 May 1940 with Hurricanes before it too moved to Northolt on 4 July. Blenheims subsequently returned during the Battle of Britain, the prelude to Operation 'Sealion', the German invasion of Britain. In order to be ready a detachment of four Blenheims from 59 Squadron arrived from Thorney Island on 11 September 1940 for anti-invasion reconnaissance flights.

When the Battle of Britain was at its height a fighter squadron returned to Hendon. 504 Squadron equipped with Hurricanes moved from Catterick on 5 September 1940, just in time for the first major raid on London on 7 September. Sgt Raymond T. Holmes took off with the rest of the squadron from Hendon on 15 September 1940 and was credited as the man who brought down the aircraft that was

Part of the wreckage of the Dornier Do17 shot down by Sgt Ray Holmes before it could bomb Buckingham Palace.

Some of the pilots and navigators of 24 Squadron, all of whom escaped from Czechoslovakia.

about to bomb Buckingham Palace, a Dornier Do17 of 1/KG76. The body of the German pilot, Hubel Gustav, was recovered from Victoria Station and buried on 25 September in Paddington Cemetery, Mill Hill.

The Hendon area had experienced bombing raids from August 1940, causing damage to local housing, but the airfield itself suffered less than many during the Battle of Britain. On the day of Gustav's funeral, however, Hendon suffered one of its worst attacks during the war. At about 2210 hours the Luftwaffe destroyed Colindale Underground station, killing more than thirteen people, four of whom were airmen from RAF Hendon: three from the Station Headquarters and one from the Radio Fitting Unit. Next day, 26 September, 504 Squadron was sent to Filton to protect the Bristol aircraft factory, which had also been attacked.

The Radio Fitting Unit is one of Hendon's more enigmatic units. It is claimed to have formed on 1 April 1940 and published sources claim that it fitted Marconi equipment in Whitleys and Wellingtons, operating until March 1941. However, another unit, the Radio Installation (Mobile) Section, left Hendon for Kemble on 18 October 1940; had the Radio Fitting Unit changed its name?

On the night of 7/8 October 1940 Hendon was attacked once more, despite the presence of anti-aircraft guns, and the hangar occupied by 24 Squadron was destroyed by an oil bomb with the loss of fourteen aircraft. On 8 November 1 Camouflage Unit arrived from Cosford, just in time for a raid that night during which high-explosive bombs hit the airfield and the hangar used by 24 Squadron's Maintenance Flight. The raid caused relatively little damage

GB2011 b/c
Geheim

Hendon (S)
Nachschublager

Karte 1:100 000 Blatt E34 | 1:63 360 Blatt 106

Kriegsaufnahme: 467 L 62

Länge (westw. Greenw.): 0° 15' 0", Breite: 51° 36' 0" (Blattmitte)
Mißweisung: -10° 55' (Mitte 1938)

Maßstab etwa 1: 15 000 (1 cm = 150 m)

Nachträge: 4.6.39.

A: G.B. 10 103 Hendon
Fliegerhorst
1) 3 moderne Flugzeughallen
2) 3 ältere Doppelhallen
3) Reparaturwerkstätten (bisher 2)
4) 4 Funkmaste (bisher 3)
5) 3 Tankstellen
6) Unterkunftsgebäude

B: G.B. 20 11 Hendon
Nachschublager
1) Flugzeughalle
2) 2 Montagehallen
3) 5 Lagerhallen
4) Unterkunftsgebäude
5) Kompensationsscheibe
6) Betriebsdepot m. Zapfstelle.

A Luftwaffe target map of Hendon, dividing it into the airfield and a store camp, despite the fact the stores had closed and some of the hangars had been demolished.

but the threat of more attacks may be the reason why 4 Motor Transport Company moved its headquarters from Hendon to Abbey Road. In order to remove some of the pressure on the Maintenance Flight a Refuelling and Rearming Party was formed on 15 November 1940 to handle visiting aircraft.

The tasks of 24 Squadron increased with the formation of 1 Aircraft Delivery Flight on 22 March 1941, using the squadron's aircraft. It was joined by 116 Squadron, which formed at Hatfield on 17 February 1941 for radar calibration duties. At first it was attached to Hendon for administration, Hatfield being a civilian airfield, but moved to Hendon on 20 April. On 12 May Hendon received two more aircraft, air ambulances donated by the Silver Thimble Fund.

Hitler postponed the planned invasion of Britain, Operation 'Sealion', but the RAF remained wary. On 10 March 1941 1416 Flight was formed at Hendon with Spitfires to watch for renewed signs of a possible German invasion; the first operational flight was made on 1 May. By then, however, Hurricanes had visited Hendon so that the pilots could practise trying to take off in a short distance before embarking in HMS *Ark Royal* in April. The aircraft were flown off and landed on Malta on 27 April in Operation 'Dunlop'. The German invasion of the Soviet Union on 22 June 1941 removed the risk of an invasion of Britain. Hurricanes of 151 Wing were sent to the Soviet Union to help stop the German advance. The Wing Headquarters moved to Hendon from Leconfield in August 1941, before sailing for Russia, arriving at Archangel on 30 August. The role of 1416 Flight changed to low-level reconnaissance over the French coast, and on 5 September it too departed, joining other photo-reconnaissance units at Benson.

1942 was a year of change for RAF Hendon. It was of little use to Fighter Command but was important as a transport hub. Changes were made with 1 Aircraft Delivery Flight moving to Croydon on 23 January followed by 116 Squadron to Heston on 20 April. The departure of some aircraft would have helped greatly because on 21 April the former 601 Squadron hangar was destroyed by fire. On 25 April, however, control passed from 11 Group Fighter Command to 44 Group Transport Command as planned. On 1 June 1942 1 Camouflage Unit moved to Stapleford Tawney followed on 9 June by the Anti-Aircraft Flight moving to Burn, replaced by 2722 Squadron RAF Regiment from Newton. The Anti-Aircraft Flight had become a separate unit on 11 August 1941. Its replacement had also been an anti-aircraft squadron, but became a rifle squadron and was more suited to the ceremonial role it might be called upon to fulfil; for example, members of the squadron provided the escort for the coffin of HRH The Duke of Kent on 28 September 1942 as it was taken from Euston Station to Windsor Castle.

On 11 December 1941 Germany had declared war on the United States following the

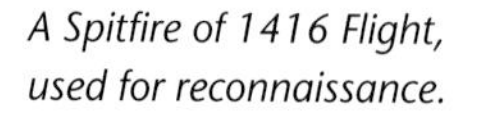

A Spitfire of 1416 Flight, used for reconnaissance.

Japanese attack on Pearl Harbor on 7 December. Almost immediately the United States began to build up its forces in the United Kingdom, and Hendon became an important hub for transport flights as Base 575. The USAAF Air Transport Command ferried aircraft across the Atlantic for the VIII Air Force in England and the XII Air Force in North Africa, but its responsibility ended at the aerial ports of arrival in Britain. Aircraft were then turned over to the VIII Air Force Service Command. Other aircraft arrived by sea and, after reassembly, had to be flown to their final destination. By July 1942 the VIII Air Force had established its own air transport service within the UK using aircraft borrowed from the RAF. In October 1942 this was designated the VIII Air Force Ferry & Transport Service. To assist with the formation of this service Major F. C. Crowley USAAF and four 2nd lieutenants were appointed to Hendon between 22 August and 17 November 1942. The USAAF used the Bellman hangars while the RAF used the rest of the airfield, but the continuing lack of accommodation forced the Americans to acquire housing away from the airfield.

The heavier transport aircraft due to enter service needed hard runways, so work began on 3 October 1942 to construct them. While they were being built 510 Squadron was formed on 15 October from A Flight of 24 Squadron, and the Maintenance Flight of 24 Squadron was expanded to form a Maintenance Wing. At about the same time WAAF personnel were selected to become Air Ambulance Orderlies. Training began in requisitioned property in Cedars Close, and they used the mess at Hendon Hall Hotel.

On 26 February 1943 the personnel and aircraft of the VIII Air Force Ferry & Transport Service were reorganised as the 2008th Transport Group (Provisional). On 15 April this became the 27th Air Transport Group with its headquarters at Meadowbank in Cranford, Middlesex, the 86th Air Transport Squadron at Hendon, and the 87th Air Transport Squadron at Warton, ferrying people, aircraft and equipment between the headquarters and depots of the VIII Air Force. The RAF also formed another transport squadron: 512 Squadron was officially formed on 18 June 1943 with an establishment of Hudsons and Dakotas, but it was not until 9 August that personnel were posted to Hendon. The air ambulance service also continued to expand with the formation of Casualty Evacuation Units: 1 Casualty Air Evacuation Unit moved to Hendon from Redhill on 28 November 1943, 2 Casualty Air Evacuation Unit arrived on 20 December, and 3 Casualty Air Evacuation Unit had moved to Hendon from West Raynham by 2 January 1944. It was disbanded into 93 (Forward) Staging Post, Lyneham, on 10 January 1944.

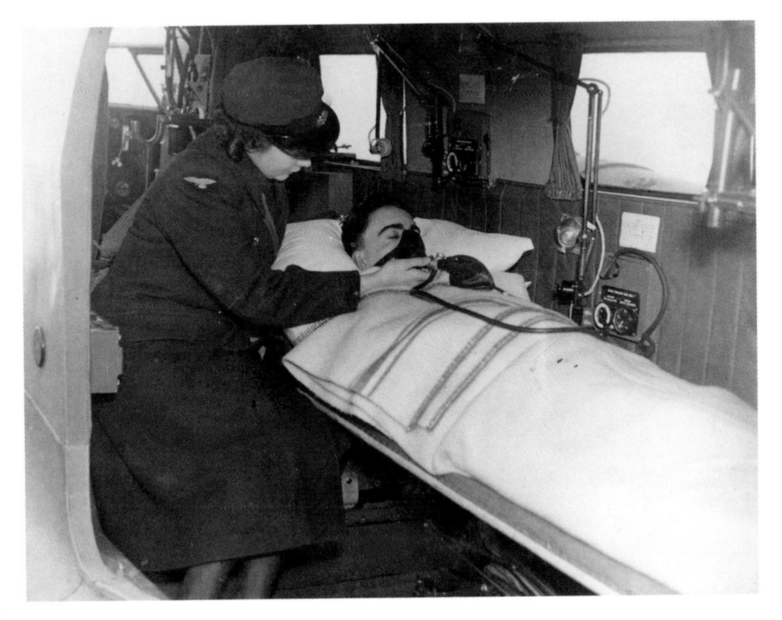

A staged publicity photograph of a nursing orderly, one of the first WAAF personnel to serve in aircraft.

The Dakotas of the 86th Air Transport Squadron left Hendon on 7 January 1944 and moved to Heston. In their place RAF Hendon expected twelve officers and 108 enlisted men. In 1943 Air Transport Command (ATC) had wished to establish its own transport operations in Europe. Most of its proposals were rejected but approval was finally given for shuttle services to operate between various bases used by the Command. Initially three C-47 aircraft were based at Hendon and flew between there and Prestwick in Scotland. By the end of the year additional aircraft allowed the operation of a twice-daily service to Prestwick, a daily service to St Mawgan and periodic flights to other stations used by ATC, a service known as the Marble Arch Line. While these operations were being developed the RAF took the opportunity to form another Dakota squadron at Hendon. A nucleus from 512 Squadron became 575 Squadron on 1 February 1944 before both squadrons moved to Broadwell later that month.

A Dakota I of 24 Squadron with the hangars and aircraft of the 86th Air Transport Squadron USAAF in the background.

On 1 January 1944 HQ 116 Wing was formed at Hendon Hall Hotel to operate internal air transport services and airline services between the United Kingdom and India, while most of Transport Command concentrated on preparations for the Normandy landings. The Dakota Conversion Section was added to the establishment of Hendon on 13 January 1944. It was originally intended that it would convert three Dakotas to passenger/freight standard and one to VIP standard per week. Instead the Section also assisted the D-Day preparations by converting aircraft to the army support role. It was estimated that 150 aircraft needed urgent conversion at the rate of twenty per week. A second Dakota Conversion Section was formed at Hendon; after training for a week it moved north in March 1944 to work at Doncaster. On 18 October 1944 the Hendon section moved to RAF Kemble.

Bombing raids in Hendon had continued throughout the war. On 7 October 1943 the air raid sirens had sounded at night. An anti-aircraft shell exploded on the airfield and subsequently killed one of the firewatchers. Possibly as a result of this the Accounts Section, Clothing Store and part of the Barracks Store were moved to four houses in Uphill Road, Mill Hill, on 1 November 1943. The author of the Hendon ORB noted the inconvenience this caused, but on 22/23 January 1944 they returned to Hendon. Other moves took place on 3 February 1944 when the WAAF accommodation moved from Cedars Close into Aeroville. This had been the home of the Book Production & Distribution Centre since 8 October 1943, one of Hendon's more

A C-47 of the American Air Transport Command.

secretive units responsible for the production of code books. Also on 3 February the keys to Uphill Cottage, 42, 46 and 56 Uphill Road, were returned to the owners' agents.

It should be noted that it was not all work and no play at Hendon. The cast of the West End play *Junior Miss* performed at the station theatre on 5 March 1944, just one of many to boost morale. In addition, Geraldo and his Orchestra broadcast the popular radio show *Workers Playtime* from RAF Hendon on 25 May 1944.

On 22 March 1944 1697 (Air Delivery Letter Service) Flight replaced the Air Despatch Letter Service of 24 Squadron before moving to Northolt on 9 May. The air service had complemented the RAF Despatch Rider Letter Service housed at the Hendon Hall Hotel. 510 Squadron was redesignated the Metropolitan Communications Squadron on 8 April 1944 and incorporated the Transport Command Communications Flight. On 12 June the Allied Flight also transferred from 24 Squadron to 510 Squadron, the Dutch element becoming 1316 Flight on 7 July.

On 24 July 1944 Hendon was transferred from 44 Group to 116 Wing, the headquarters moving from the Hendon Hall Hotel to the former Auxiliary Air Force Mess, much to the annoyance of all at Hendon. The Metropolitan Communications Squadron flew routes in the United Kingdom while 24 Squadron concentrated on the longer routes to India. As Europe was liberated 107 Wing was formed at the Hendon Hall Hotel on 22 October before moving overseas to control transport routes across France. The American Air Transport Command (ATC) service was also expanding and adopting aircraft too large for Hendon; as a result it was necessary to move the headquarters to Bovingdon in Hertfordshire. The fifty Dakotas of the ATC left for Bovingdon on 16 October 1944 followed by the ground staff on 19 October. It is unconfirmed, but this may have been 1402 AAF Base Unit; there is reference to this unit moving to Bovingdon by 31 October 1944.

German V-bomb attacks on Britain began almost as soon as the Allies had landed in Normandy, the V-1 flying bombs launched from sites in the Pas de Calais. Four members of the Women's Auxiliary Air Force were killed by one that hit Colindale Hospital on 1 July 1944, destroying two wards. On 25 July

RAF Hendon in 1943, showing the runways and aircraft dispersed around the perimeter.

A Dakota at Hendon in June 1944. This is the first photograph to show the Grahame-White offices being used as flying control.

another landed in Sunny Hill Park and caused damage over a wide area. On 3 August another flying bomb destroyed a brick-built barrack block on RAF Hendon, killing nine and injuring twenty-five. RAF personnel had to be assisted by the USAAF and soldiers from the Army School of Physical Training billeted at the Police College. The various raids had kept the men of 6224 Bomb Disposal Squadron busy; it was later reduced to a Flight as part of 5134 Bomb Disposal Squadron.

On 1 January 1945 116 Wing was disbanded into 47 Group at Hendon. The remaining months of the war saw the clearance of the remaining debris and the gradual rebuilding of Hendon. Despite the conditions it continued to host royalty, politicians and military commanders, flying to or from the airfield. For example, on 17 July 1945 three Dakotas bearing Their Majesties The King and Queen, HRH Princess Elizabeth, their entourage and accompanying press corps flew from Hendon to RAF Long Kesh for the first Royal Visit to Northern Ireland by air. The Air Despatch & Reception Unit received passengers, freight and mail from around the world, but Victory in Europe did not bring a reduction in aircraft movements; indeed, it saw an increase, with a peak of 3,897 RAF aircraft movements and 410 USAAF and USN in July 1945. Victory did at least allow for a VE dance in the NAAFI on 8 May, a service of thanksgiving in the station theatre on 13 May and participation in the victory parade in Hendon Park on the same day. On 16 May Ralph Reader brought the RAF 'Gang Show' to Hendon. Unfortunately the party atmosphere was spoiled the next day when the hangar of C Flight Metropolitan Communications Squadron caught fire, resulting in considerable damage and the death of two members of the squadron.

Aircrew of the Metropolitan Communications Squadron in front of an Airspeed Oxford, September 1944.

Peace Returns 10

Hendon had held one of the last of the pre-war flying displays in 1939, and on 15 September 1945 it held one of the first. More than twenty aircraft took part in the flying display watched by tens of thousands of spectators. There were displays too in many of the buildings, showing wartime developments that had been secret, including radar. It was estimated that between 60,000 and 100,000 attended the display.

On 25 October 1945 the station and its units were transferred to 46 Group Transport Command. In November 1945 Hillfoot and Canberra, which had been used as WAAF officers' quarters and mess were vacated due to structural defects. Hill Croft and Yew Ridge in Cedars Close, which had been officers' quarters, were also finally vacated in December 1945. The station was relieved when 47 Group began to leave Hendon in May 1946 and handed back misappropriated buildings.

The same thing happened at the end of the Second World War as at the end of the First. Newspapers reported that RAF Hendon was to close and become an estate of approximately 1,000 houses. This created a lot of local speculation, which was reported in the station ORB. In February 1946 it was hoped that the statement by the Air Ministry that Hendon would continue to be used for flying would end the speculation. Hendon's future was confirmed by the reconstruction work undertaken by the men of the Airfield Construction Flights in the late 1940s, most buildings requiring some corrective work. Canberra House, once defects were rectified, was put back into service as the station commandant's house and the surviving Grahame-White factory buildings were modified for use as the passenger and freight terminal.

Although Hendon continued as an airfield there was a reduction in the number of units housed. 24 Squadron moved to Bassingbourn between 14 and 25 February 1946 together with 4024 Servicing Echelon; the latter had formed on 26 July 1944 to service the squadron's aircraft. The Coastal Command Communications Squadron moved in from Leavesden between 25 February and 1 March 1946 before it disbanded on 1 May 1946. 1316 Flight disbanded on 4 March 1946. The Fighter Command Communications Flight

A Percival Proctor of 31 Squadron in 1951.

was another temporary resident, moving in on 27 April 1947 before moving to Bovingdon on 9 July. On 19 July 1948 the Metropolitan Communications Squadron became 31 Squadron before reverting to its previous name on 1 March 1955.

The RAF continued to share Hendon with its American colleagues. Although no USAF units were stationed there after 1944, it continued to be an important destination for them. From 1946 Hendon housed the USAF Air Attaché's flight until it moved to Bovingdon in December 1949. In 1948 Spitfire PL983 was allocated for the use of the Civil Air Attaché, Livingston Satterthwaite. Before he could fly it, however, Lettice Curtis, the former ATA pilot, had to fly it to an airfield with longer runways.

The absence of the USAAF enabled the formation of a US Navy transport unit at Hendon in 1946. During the first half of 1942 the US Navy had commenced a thrice-weekly service between Hendon and Eglinton near Londonderry, stopping at Long Kesh to offload and collect passengers, light freight and mail, using Lockheed 12 and R4D (the US Navy version of the Dakota) aircraft. During 1946 it was determined that a Utility Transport Squadron was required to support US Navy forces in Europe, and on 3 December 1946 squadron VRU-4 was commissioned at Hendon with five R4Ds and four JRB aircraft. Like the USAAF it used the three Bellman hangars and housing around North West London for its personnel.

VRU-4 operated over all of Europe, Scandinavia and North Africa with a detachment at Port Lyautey in French Morocco, where major servicing was undertaken. On 10

A Douglas R4D of VR-24 US Navy, circa 1950.

December 1946 a second detachment was formed at Naples, but that ceased in April 1947 and the aircraft transferred to Port Lyautey. In July 1947 two more R4D aircraft were attached to the squadron, but soon one of these and a JRB had to be attached to Port Lyautey. On 1 September 1948 the squadron designation changed to Transport Squadron VR-24. Two R5D aircraft of VR-1 had been attached to the detachment at Port Lyautey, but the squadron received its own in October 1949. Hendon was too small, so the aircraft had to be based at Heathrow.

Although Hendon was a station in Transport Command, the AuxAF returned once more. Reserve Command was formed on 7 January 1946 to administer AuxAF and RAF Volunteer Reserve units when they reformed, the Air Training Corps and University Air Squadrons. The AuxAF reformed on 10 May 1946 and so too did the squadrons, with 601 and 604 Squadrons back at Hendon and recruiting once more. On 23 August 1946 Headquarters 65 (London Reserve) Group formed at Hendon to administer the AuxAF squadrons in the London area, and in 1947 King George VI granted permission for the AuxAF to be renamed the Royal Auxiliary Air Force. A further administrative change came on 1 August 1948 when Reserve Command was renamed Home Command and 65 Group dropped 'London Reserve' from its title. Hendon was only just large enough for Spitfires, and too small for jet fighters, so when in 1949 the squadrons were due to re-equip with Meteors they had to leave once more. Both squadrons moved to North Weald, 601 on 27 March and 604 on the following day. On 1 February 1951 65 Group disbanded at RAF Hendon, its headquarters having amalgamated with 61 Group at Kenley.

Another Auxiliary unit formed at Hendon was 1958 Auxiliary Air Observation Post (AOP) Flight, part of 661 Auxiliary AOP Squadron. The squadron was formed on 1 May 1949, as were other flights at Kenley and Henlow. Although it was part of the RAF, many of the volunteers were from the Territorial Army and it worked closely with the Royal Artillery units of the British Army's Eastern Command. The squadron and its flights disbanded on 10 March 1957, together with all the other squadrons of the Royal Auxiliary Air Force. 1958 Flight was not the first Hendon unit to be equipped with Austers; the RAF Antarctic Flight had been formed at Hendon on 25 April 1949, consisting of two officers, a sergeant and two corporals, and took two Auster AOP.VI aircraft (VX126 and VX127) to the Antarctic on a joint British-Scandinavian expedition. The aircraft were flown on both floats and skis, helping the expedition to reach its destination on Queen Maud Land. Once its work was done the flight returned to Hendon and disbanded in January 1951.

The 1950s were relatively quiet years for Hendon. On 20 March 1950 it ceased to be controlled by 46 Group and came directly under Headquarters Transport Command. It also became a temporary home to 604 Squadron once more. The Esher Trophy was presented to the squadron at Buckingham Palace on 21 June 1950, but the full rehearsal for the parade was held at Hendon. On 21 August a conference to discuss the civilianisation of servicing and repair of aircraft and mechanical transport was held as a way to reduce manpower and costs. Even at the end of 1950 rebuilding was being undertaken, one

Spitfires of 601 Squadron, 1949.

An Auster AOP.VI of 661 Squadron.

aim being the move of WRAF officers from 18 Avenue Road. In February 1951 a Passenger Handling Section was formed at RAF Ruislip, but on 8 October it moved to Hendon and on the following day reformed as the Air Trooping & Freight Control Section. Its function was to deal with the assembly, disposal and transit of passengers travelling under the Air Trooping Schemes and was the forerunner of the Joint Services Air Trooping Centre (JSATC).

US Navy operations had been growing in the Mediterranean and the detachment at Port Lyautey acquired greater importance. As a result, on 1 August 1950 the headquarters of VR-24 transferred there and Hendon became a detachment. By 1 June 1951 the Hendon detachment had the R5D, three R4D and two JRB aircraft. On 24 June 1952 the VR-24 detachment was replaced by the formation of VR-25, with a detachment at Naples. In 1953 the squadron at Hendon was redesignated Fleet Aircraft Service Squadron FASRON 76, and the Naples detachment was redesignated FASRON 77. In about 1955 FASRON 76 was renamed again as FASRON (Special) 200. The US Navy was not the only organisation to keep changing unit designations; as already mentioned, on 1 March 1955 31 Squadron became the Metropolitan Communications Squadron once more.

Some of the exhibits at the Daily Express Exhibition of 1951, marking fifty years of the Royal Aero Club.

King George VI had known Hendon for many years, having attended various displays and used aircraft based there. His death in February 1952 led to a period of mourning that was observed until 31 May. When the news of his death was received, all sports events and official social functions were cancelled, places of entertainment on the station closed and only essential flying continued. A memorial service was held on 15 February and an officer, flight sergeant and three airmen attended his funeral.

Hendon's role as a transport hub meant that it was often called upon to help with disasters. In February 1953 the station took part in Operation 'King Canute'. Winter storms had caused flooding that affected much of the East Coast of England and the Netherlands. From 1 until 19 February personnel from the station helped repair sea defences at Tilbury, Purfleet and Canvey Island. Then in November 1956 the RAF Hendon ORB records that the 'intervention situation put extra tasks on the Air Trooping Centre'. The intervention was the Anglo-French invasion of Egypt, and in addition to the extra movements created by a major military operation the centre also had too look after wives and children; between November 1956 and January 1957 families were evacuated from Egypt and Libya, and approximately 1,300 passed through Hendon.

Buildings at the entry into the US Navy facility at Hendon.

In September 1956 FASRON (Special) 200 left Hendon for Blackbushe with its R4D aircraft. The liaison flight of the 32nd Anti-Aircraft Artillery Brigade of the US Army moved in for a few months but they too departed in 1957, marking the end of a fifteen year association between the Americans and the station. A conference held on 17 June 1957 paved the way for the creation of the JSATC by bringing in the Army and Navy. On 15 November the station transferred from Transport Command to Home Command and a new station commander was appointed. The JSATC was to form as the 'Commanding Unit' with responsibility for the RAF Design &

A Devon C.2 of the Metropolitan Communications Squadron flying over Hendon before departing on 4 November 1957.

Display Unit (DDU), a Civilian Aerial Erection Flight and the historic aircraft collection of the Royal Aeronautical Society. By then, however, the Metropolitan Communications Squadron had departed to Northolt on 4 November, taking with it the Transport Command Communications Flight, and Hendon no longer housed any flying units.

The most northerly Bellman hangar (Building 106) was used from November 1957 for the storage of the Nash Collection of aircraft on behalf of the Royal Aeronautical Society. In May 1957, however, it had been announced that the Air Council had in principle approved the use of the Grahame-White hangar as an exhibition hall for the collection. Many of these aircraft had been seen at Hendon displays in the 1930s and included one of Major Savage's SE.5a aircraft. They were joined in February 1958 by fifty-eight boxes of exhibits from 15MU at Wroughton on behalf of the Air Historical Branch. It was a temporary arrangement, and in November 1959 the aircraft moved to British European Airways at Heathrow; they have since returned as part of the RAF Museum Collection. The same month saw the arrival of the Design & Display Unit from RAF Stanmore.

On 1 April 1958, the 40th anniversary of the formation of the Royal Air Force, the Joint Services Air Trooping Centre was officially formed and took over the task of running Hendon. Now that it no longer housed flying units, 90 Group began the task of removing redundant radio equipment. As well as removing it from air traffic control and station headquarters, 90 Group also had to dismantle a VHF transmitter/receiver station at Scratchwood, and a VHF direction-finding station at Copthall. The task was completed by 30 January 1959.

It is ironic that while the air traffic control equipment was being removed some flying resumed at Hendon. Gliding had first taken place in 1950 with the move of 142 Gliding School from North Weald to Hendon by December of that year. It stayed at Hendon until it moved to Hornchurch on 1 August 1953. In June 1958 approval was given for the formation of 617 Gliding School at Hendon on a six-month trial. The school was formed on 25 November 1958 and gliding started on 11 January 1959; the six-month trial lasted until 1968. By the time it started the station had been transferred once more and was now in Maintenance Command.

The absence of flying units allowed Hendon to house a wide range of units and provide accommodation for others. For example, the Central Depository was planned to move from Titchfield into Hangar 66 in 1958, while the RAF Cinema Corporation used the former

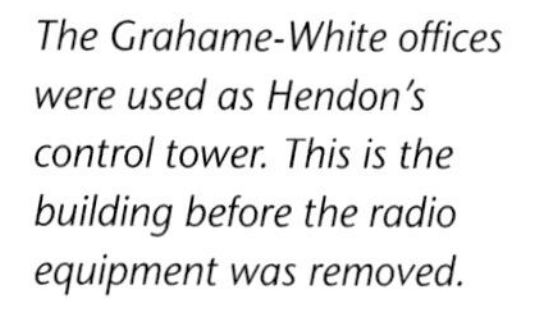
The Grahame-White offices were used as Hendon's control tower. This is the building before the radio equipment was removed.

Decontamination Centre as a store and the Meteorological Office was given permission to use the Grahame-White hangar as storage. The airfield was used for rallies, exhibitions and safe driving trials. Technical Training Command used the airfield for David Brown tractor driving courses over two months, and the Road Research Laboratories used it to test road signs. In July 1958, however, the Commanding Officer requested new accommodation at Hendon. The work needed included the construction of a new barrack block for permanent staff, a new permanent airmen's mess and dining room together with a new NAAFI. New barrack blocks were needed for use by transient men of all three services because they were still housed in the wartime huts.

On 17 October 1960 705 Signals Unit moved to Hendon, administered by Headquarters 1 Group RAF Bomber Command. It provided radar bomb scoring facilities and in its first month recorded 229 simulated RAF and USAF bombing raids on London. On 7 July 1961 the unit received a new plotting vehicle and appears to have left Hendon by September 1965. Op Order 1/64 provided a Defence Plan for RAF Hendon. People do not think of Hendon as a military target, but it was one of the targets for the simulated bombing raids on London and it is probable that the Soviets too would have targeted the area. The security plan, however, also had to take insurrection into account. This type of defence became more important in the 1970s as the threat from terrorism increased.

Computers are things we now take for granted – much of this was written using a laptop computer. In the early days, however, they were much larger, needing their own buildings to house them. RAF Hendon was chosen as the new home for the supply control computer of Maintenance Command, and on 1 July 1961 the Supply Control Centre (SCC) was formed in Buildings 59 and 60. By 4 June 1962 it was able to move into its own purpose-built building, and on the 30th the RAF ensign was flown from the flagstaff outside SCC for the first time. The first system came into use in January 1966 and its official opening took place on 28 June. By 1969 approximately 530,000 items were under computer control out of a total of 900,000. In 1969 an upgraded automated data processing system was planned, which could handle all the stores for Army and Navy aircraft as well as the whole of the RAF inventory.

Gliders of 617 Volunteer Gliding School waiting to launch.

The monotony of daily life on a station was often broken by parties, fetes, fairs and displays. Hendon, however, managed to go one better in July 1966. On the 10th shooting started for the film *The Dirty Dozen*; it was completed by 13 July and the site was cleared by the 15th. The guardhouse at Hendon features clearly in the film and, despite its brief period of fame, the event is still a talking point in the area.

In May 1967 Hendon began its contribution to the largest transport operation since the Berlin Airlift, when all British service personnel

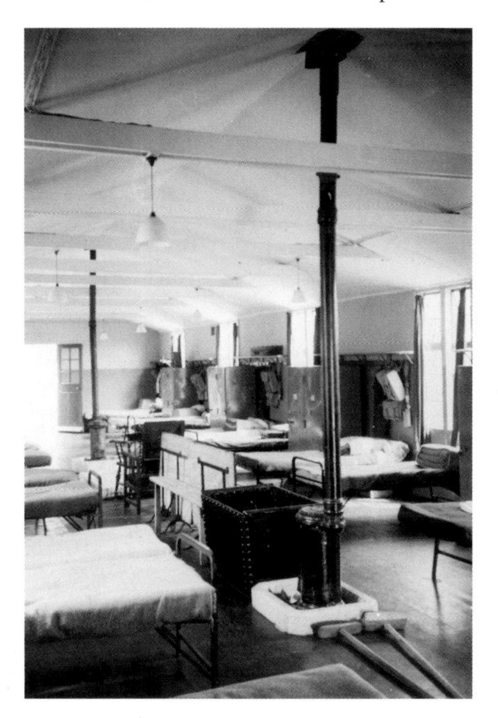

The interior of one of Hendon's wooden barrack huts in 1951.

The station flagstaff outside the SCC building.

On 7 June 1963 the married quarters were flooded to a depth of 2 feet by a cloudburst. This area is not known for flooding, but the same happened again on 18-19 November.

The AOC's inspection was an annual event for every RAF station.

Aircraft on display during the Founder Members' Day, 5 July 1968.

and their families had to be evacuated from Aden at short notice. A JSATC detachment was sent to Gatwick Airport to work with the Soldiers, Sailors, Airmen & Families Association. Its work involved the reception of the families who had been evacuated and the provision of the necessary assistance. The unit's work was completed by July, and the final military personnel were evacuated by November 1967.

The airfield was now redundant, and as early as 10 May 1966 a meeting had been held at Hendon regarding the development of part it for housing. Despite its closure, aircraft did occasionally use it by accident. On 2 June 1966 a C-54 transport of the United States Air Force landed following an in-flight fire warning light. Two port tyres burst on landing, but it was able to depart the next day after repairs. On 25 January 1967 a Nord Noratlas of the West German Air Force mistook Hendon for Northolt and landed by accident.

The gliding school ceased flying on 31 March 1968, and on 1 April it moved to Bovingdon. Before the airfield was redeveloped, however, it saw one last arrival and hosted one last display. On 19 June the last aircraft planned to use the airfield's runways, a Blackburn Beverley, arrived and became gate

The arrival of the Blackburn Beverley on 19 June 1968 resulted in numerous calls to the emergency services by worried neighbours.

Grahame Park Way begins to take shape in the spring of 1970.

guardian for RAF Hendon. 1968 saw the 50th anniversary of the formation of the RAF, and on 5 July a small display was held to celebrate the occasion, the Royal Air Force Founder Members' Day.

On 28 October 1968 the airfield was taken over by the London Borough of Barnet, with an official handover ceremony in the officers' mess the following day. Before construction covered the site, however, there were two more landings. On 22 December Piper Cherokee G-AVWD landed, lost in bad weather. On the next day Cherokee G-AVUR landed with a pilot to collect the plane, and both departed to Blackbushe, assistance being provided by RAF Northolt. This was the last fixed-wing landing at Hendon, but it might not have been. Between 4 and 11 May 1969 Hendon was a diversion airfield for the Harrier in the *Daily Mail* Transatlantic race, although it was not needed.

Construction of the Grahame Park estate began in 1969. More than 1,700 homes were built on the aerodrome for the Greater London Council, with most of the design work undertaken by the late Michael Brown, and the first were occupied in 1971. Aerodrome Road was used for the construction traffic, but removal of the station main gate allowed the public to start using it, despite its condition. Gates were still maintained near the railway bridges and closed one day each year to remind the public that the road was private; this continued until the road was adopted by Barnet Council in the 1980s and metalled.

Hendon remained a busy RAF station accommodating the SCC and JSATC. It was unusual because it was split into two parts by the housing estate. West Camp housed most of the accommodation, including the Transit Hotel, while East Camp housed the SCC and various others, including 120 Squadron, Air Training Corps. The 1970s and 1980s saw little change at RAF Hendon – most was on the site of the Royal Air Force Museum. Further upgrading of the SCC system saw the inauguration of a new ICL 4-72 computer at SCC on 30 June 1975. This was followed on 5 March 1976 by a cocktail party to mark the cessation of the old AEI 1010 computer.

On 22 January 1974 the Transit Hotel was renamed Airbridge House.

The RAF Museum

11

While major changes were planned for Hendon in the 1960s, they did not affect the residents much at first. Typical of the small changes that did happen concerned the use of buildings. Between 4 and 8 June 1962 Education moved into Building 59, and Building 60 was used as overflow storage for the station equipment officer once they had been vacated by the Supply Control Centre. They remained there until they had to make way for the RAF Museum.

John Tanner, Librarian at the RAF College, was not the first to suggest the creation of a Royal Air Force Museum. His ideas were taken seriously, however, and a search was made for a suitable site. The RAF Museum began acquiring objects in 1965, which were assembled at its first store at RAF Henlow. A more suitable location for a museum was required and on 2 August 1965 John Tanner inspected the Grahame White hangar as a possible location. On 24 November 1966 Air Marshal Sir Charles Broughton, Air Member for Supply & Organisation, also came to Hendon to look at the site of the proposed museum.

Plans changed and the museum was allocated the two remaining First World War hangars to form the core of a new building. On 29 December 1967 John Tanner took over Buildings 59 and 90D (Grahame-White) on behalf of the RAF Museum until the planned construction was completed, after which aircraft of the AHB Collection and the Nash Collection returned to Hendon, where they were joined by a number presented by Hawker-Siddeley. Gallery displays, constructed with the help of the resident Exhibition Design & Display Unit, complemented the aircraft and told the history of the RAF.

The RAF Museum was opened on 15 November 1972 by Her Majesty Queen Elizabeth II in what are now known as The Historic Hangars, the two remaining General Service Aircraft Acceptance Park coupled hangars. The roof structure consists of Belfast truss latticework, which gives this type of hangar its common name of a Belfast hangar. The two hangars differ in that the northern one was built using concrete for the central roof supports while the other uses wood. There were large wooden doors at each end of the hangars, parts of which can be seen on either side of the entrances to the Bomber Command Hall and Dermot Boyle Wing. The latter was the first extension, opened by HRH Prince Charles on 6 December 1975 as a space for temporary exhibitions. The first of these was the 'Wings of the Eagle' exhibition, which brought to Hendon many of the aircraft that would feature in the Battle of Britain Museum.

The Battle of Britain Museum was opened in 1978 by HM Queen Elizabeth the Queen Mother, and is the only one that does not use or occupy the site of any former buildings. It was originally designed to resemble a hangar from the Second World War, but was given a new external skin and a glass wall in 2009. Further expansion took place in 1983 when the Queen Mother opened the Bomber Command Museum. Its construction required the demolition of several structures built in 1917 for the Acceptance Park.

By the 1980s Hendon no longer had a place in RAF plans. Most of the RAF trooping flights were undertaken by commercial airlines, and Airbridge House was redundant. Computers too were changing, and the SCC began to move to RAF Stanbridge. The station's last commanding officer, Wing Commander William George Simpson, welcomed the Queen Mother to the official closure ceremony on 1 April 1987. Bad weather forced the parade to be held in the Grahame-White hangar, and most of the planned flypast was cancelled.

The closure of Hendon took some time to complete and RAF personnel were still present in 1988. Construction of the 'bird' estate on West Camp, so called because the roads and apartment buildings are named after birds, began almost immediately. East Camp remained abandoned; most of the remaining buildings were demolished in 1993, except for the Grahame-White factory buildings, the former officers' mess and a few other buildings from the 1930s. In 1994 the first development on East Camp started; the officers' mess was acquired by Middlesex University as a hall of residence, Writtle House, and Platt Hall was constructed in its grounds. This was followed by the construction of Colindale Police Station, which opened on 10 March 1997. The museum acquired the remaining RAF buildings adjacent to its site.

When the RAF Museum was formed it was part of the Ministry of Defence, with a Board of Trustees to oversee its running. The other two museums on the site, however, were separate from it and had their own Boards of Trustees. In practice, however, the same people served on all three Boards. Funding also differed, the RAF Museum being funded by the Ministry of Defence, while the Battle of Britain and Bomber Command Museums received no government money and were dependent on admission charges and fund-raising. In 1984 the status of the RAF Museum changed because of an enabling clause in the National Heritage Act. It became a Non-Departmental Government Body (Quango), and its employees were no longer civil servants but employees of the Trustees; in practice few noticed the change.

A shortfall in income for the Bomber Command Museum placed it heavily in debt and threatened its existence; drastic action was required, resulting in the merger of the three museums in 1989. The Ministry of Defence paid the debts of the Bomber Command Museum but required the Trustees to repay the money through reduced central funding, forcing the museum to impose an admission charge across the site. Even though the debt was repaid, funding never recovered. In December 2001, however, the museum was able to withdraw the entry fee after the MOD agreed to an increase in its grant, matching the national museums parented by the Department of Culture, Media & Sport.

On 17 December 2003 the centenary of heavier-than-air powered flight was marked. Recognising the significance of the event, planning began for the opening of another exhibition hall. With the aid of National Lottery funding, construction began of the Milestones of Flight Exhibition, opened by Prince Phillip, Duke of Edinburgh. Unfortunately the construction required the demolition once more of existing buildings, including Building 59, which had been built in 1917, had been used as offices by 601 Squadron, had housed the Supply Control Centre, and was one of the first to be occupied by the museum.

The land acquired in 1994 allowed for another building to be constructed in the museum grounds while Milestones was prepared, and uses were found for two of the buildings on the site. Building 69 was the

Parachute Store and had housed the Civilian Aerial Erection Flight. It was restored and leased to a computer company, but now provides offices for museum staff. Building 51 (Drg 814/30) was workshops for the AuxAF squadrons, but subsequently became home to the Design & Display Unit, later named the Exhibition Production Flight (EPF). It was restored as the museum's workshop in place of one demolished for the construction of Milestones. Another, Building 52 (Drg 840/30), was the stores for the AuxAF squadrons before

***Above left:** John Tanner, Sir Dermot Boyle and the Duke of Edinburgh examining a model of the RAF Museum in the officers' mess. On 4 June 1970 the Duke of Edinburgh paid a visit as Patron to see progress.*

***Above:** The opening of the RAF Museum on 15 November 1972 by Her Majesty the Queen.*

***Left:** The opening of the Bomber Command Museum in 1983 by HM the Queen Mother; she had previously opened the Battle of Britain Museum.*

The Duke of Edinburgh speaks at the dinner to mark the 25th anniversary of the opening of the RAF Museum.

becoming the station stores, otherwise known as Supply Control & Accounting Flight (SCAF). At present it is empty and awaiting restoration.

The fourth building on the site was Building 46, Vickers Block (Drg 774/30), built in 1929 or 1930 as a combined dining room, institute and sergeants' mess. Unfortunately it was damaged by contractors in 1994 and subsequently demolished. This, however, allowed the developers of East Camp to move the Grahame-White hangar onto the museum site in time for the opening of Milestones of Flight and the construction of housing on the former East Camp, now known as Beaufort Park.

Construction of the Beaufort Park estate began in 2005. On 12 July 2006, as the first flats were nearing completion, the buildings were burned down in a few minutes. Platt Hall also suffered serious damage, but the Grahame-White Watch Office survived. This event, together with similar fires elsewhere, prompted a review into the safety of wooden-framed apartment buildings. A different construction method was used in its reconstruction and subsequent apartment buildings, with the development winning several awards.

The Grahame-White hangar had been separated from the Grahame-White Watch Office during the Second World. The

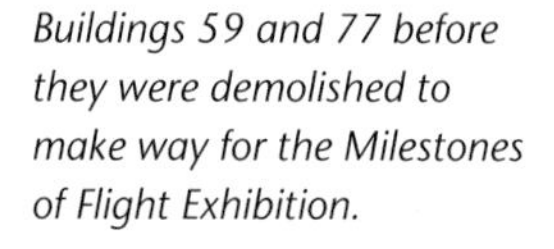

Buildings 59 and 77 before they were demolished to make way for the Milestones of Flight Exhibition.

Vickers Block. This building was badly damaged in 1994 and demolished to make way for the Grahame-White hangar.

developers also wanted to move this historic listed building, but English Heritage was reluctant until it was proved that no one could be found to occupy it in its original location. It therefore agreed to the move, and in 2010 the Watch Office was rebuilt next to the Grahame-White hangar and linked to it for the first time in more than fifty years. Eventually these buildings will form a new exhibition showing the history of the aerodrome, aircraft construction in North West London, and the history of the Royal Flying Corps and Royal Naval Air Service.

Construction is continuing in Beaufort Park, but it is not the only development in the area. New Hendon Village has been constructed on previously open ground in the centre of the Grahame Park estate, allowing the Council to move residents from the original flats, which are being demolished and will be replaced. This will be one of the biggest redevelopments in Europe and will take about fifteen years to complete. Colindale tube station is also being refurbished so that it can cope with more commuters, not just from Grahame Park and Beaufort Park but also a new development on the site of Colindale Hospital. Further developments are expected on the site of the Newspaper Library and Peel Centre as part of the Colindale Area Action Plan.

The Grahame-Park estate in the 1980s; work has started to replace it with a major new development.

POSTSCRIPT

The Grahame-Whites after Hendon

Contrary to popular belief Claude Grahame-White did not turn his back on Britain and aviation. He may have turned to property development as his main source of income, but he remained a member of the Royal Aero Club and owned property in England.

In 1925 Grahame-White became agent for Baby Gar speed boats. The American Garfield 'Gar' Arthur Wood was a successful inventor and businessman who held the world water speed record on five occasions and dominated the Harmsworth Trophy. Alfred Charles William Harmsworth (later Lord Northcliffe) had presented the trophy for countries to compete against each other in high-speed motor boats. The first race for the trophy was held in 1903 and was won by Dorothy Levitt. When the races restarted after the First World War they were dominated by Gar Wood and his son, despite the best efforts of many including Henry Segrave and Malcolm Campbell.

Gee Whiz was one of Claude's Baby Gars, fitted with a Liberty aero engine. He first used it in 1925, winning the Grahame-White Challenge Trophy at the British Motor Boat Club regatta at Hythe. At Bournemouth in August 1927 he used it to win the Holbrook Cup outright. He had another Baby Gar at that meeting called *Attaboy*. He also used *Gee Whiz* at the Hythe meeting in August 1927, winning his own trophy outright. Other boats he had at Hythe were *Cutie* and *TNT*; he raced the latter at Hythe in July 1926,

Mrs Grahame-White (Ethel Levey), Senatore Guillermo Marconi, Mrs Walters and Claude Grahame-White at Seaview on the Isle of Wight in 1925, with the Chain Pier in the background.

but we do not know if they too were Baby Gars.

In late 1926 or early 1927 Claude and Ethel sold the yacht *Peter Pan* to Lord Louis Mountbatten and his wife. In her place, however, he bought the steam yacht *Ethleen* (485 tons) and took Ethel on an extended cruise of the Mediterranean. He also took with him at least one Baby Gar, which he raced, and the success of the craft prompted him to open a showroom in Cannes. In 1926 he also bought the *Lady Vagrant* (485 tons) of 1903 from Charles Markham. She was one of the yachts present at the 1937 Coronation Review at Spithead, and Claude entertained his guests on board. She was requisitioned in August 1939 but was released from service in February 1940, the Admiralty taking Montague's yacht *Majesta* instead.

In September 1927 Claude sailed from Southampton to New York on the *Aquitania*, returning to Southampton on 2 November on the same ship. This marks the start of a period when Claude and Ethel spent most of their time in the United States. Claude bought Balfour Cottage, Balfour Place, London W1, and Miraflores, Palm Beach, as private residences. Ethel continued her acting career, seeking roles in Hollywood, and Claude moved into real estate.

On 4 July 1931 Claude and Ethel returned to Plymouth on the *Belgenland*. Once back in England, on 10 July he placed Brytcast Stainless Metals Ltd in voluntary liquidation. The company had been registered in 1930 to acquire the steel manufacture and moulding business of Martin Industries Ltd, Blackheath, Birmingham, but presumably failed. In 1933 the Grahame-Whites sailed once more to New York. Claude moved down to Palm Beach but Ethel stayed in New York, still trying to find suitable roles in the theatre, but unfortunately her age was against her. When Claude returned to England in 1934 Ethel stayed in America.

In 1934 Claude had Rossmore Court built. He kept one of the flats, which remained the registered address of the Grahame-White Co Ltd

Grahame-White's yacht Lady Vagrant. *This illustration is from a souvenir produced by the company that chartered her for the 1937 Spithead Coronation Review.*

Garfield Wood's Baby Gar IV*, similar to the craft sold by Claude Grahame-White.*

until his death. When Ethel did return to the UK it was to start divorce proceedings, despite attempts by Claude to persuade her to rejoin him. They divorced in 1939, and Claude married his third wife, Phoebe Lee. Ethel died in New York on 27 February 1955, aged 73, but Phoebe survived Claude and donated his papers to the RAF Museum before her death in Sussex in 1993.

Although he had moved into property development, Claude did not abandon boats. He was a director of Aero-Marine Engines Ltd, which acquired the rights to the Lorraine aero engine and converted it for marine use. Four were fitted in two pairs in the Aero-Marine 40K, which was designed as a motor torpedo boat. Attempts in 1939 to interest the French and British navies failed, but eleven were sold to the Republican Spanish Navy. One was built at Looe in Cornwall and the others in French yards. Before they could be delivered, however, nine were captured and used as air-sea rescue boats by the German Navy. Aero-Marine Engines Ltd was finally struck off the company register in 1968.

In 1932 Claude became a regular weekend visitor at Heston and had his first flight for ten years with Captain Valentine Henry Baker, one of the founders of Martin-Baker Aircraft Co Ltd. He was also honorary treasurer of the British Gliding Association, holding the post between 1933 and 1934. His obituary in *Sailplane and Gliding* for October 1959 reveals why he stopped flying; he is reported to have told the Council that 'he was prepared to do anything so long as he was not asked to fly, as his wife forbade it.' He soon had to relinquish the post due to other commitments, but was able to make a donation to the association's funds. In 1937 Mr Howell won the Grahame-White Prize at the Eastbourne Flying Club's 'At Home' day in September 1937 at Wilmington aerodrome. Unfortunately I do not know if he maintained links with the Bedford & District Model Aeroplane Club of which he was president before the First World War.

When war broke out Claude offered his services to the Government once more, but now his age was against him. He was restricted to being a fire watcher while he and Phoebe divided their time between Cowes and Rossmore Court. After the war he helped rebuild London, which enabled him to establish a new home at Windlesham and later at Roquebrune near Monte Carlo. He was on his way to Roquebrune in December 1953 when he had his first serious motoring accident after fifty-seven years of driving, putting him and his wife in hospital. In 1951 he returned to Hendon for the Daily Express Exhibition, and was also a guest at a Royal Aero Club dinner on 26 November 1957 for pioneers. On 30 April 1959 he was at the Royal Aero Club once more at a dinner to mark the 50th anniversary of the first flight in England by Lord Brabazon, probably his last public appearance.

Claude died in Nice on 19 August 1959, aged 80. A memorial service was held on 16 September at Christ Church, Down Street,

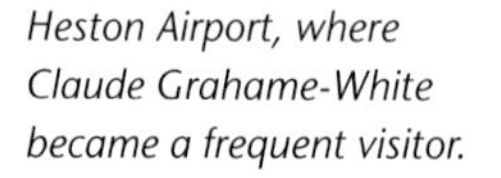

Heston Airport, where Claude Grahame-White became a frequent visitor.

London, and there is a memorial to him in Golders Green Crematorium. Phoebe Grahame-White presented the Grahame-White Challenge Trophy to the Royal Aero Club, which in 1960 awarded it to the winner of the Cardiff-London air race. In 1966 the company that started as Grahame-White, Bleriot & Maxim Ltd was finally wound up, and Claude's other companies, Portman Trust Ltd, Gar Wood Motor Boats Ltd, London General Garage Co Ltd, Hotel Trust Ltd and Clapham Properties Ltd followed. Different authors have made many claims about Claude Grahame-White, some of which are contradictory, but it is obvious that he was an excellent communicator and a showman who did much to promote Hendon and aviation in general.

Of the three siblings, the least is known about Beatrice. She married Captain James Seton Drummond Gage on 20 August 1927 at Holy Trinity Church, Cowes, and Claude hosted the wedding reception on his yacht *Ethleen*. Nothing more is heard of her until she died a widow at Faygate on 27 January 1953.

Montague survived his brother. On 23 December 1912 he was commissioned as a Lieutenant in the Royal Naval Volunteer Reserve and spent most of the war crossing the Channel in motor yachts converted to act as hospital ships. On 1 July 1918 he married Annie Louise Watkins, the former 'Birdie Sutherland', and in 1920 was promoted to the rank of Lieutenant Commander RNVR; he is often referred to as Commander Grahame-White. After the war he became a successful car, yacht and aeroplane broker, his customers including the aristocracy, businessmen and celebrities. He had owned Warsash House before moving into its Clock Tower due to financial difficulties, selling the house and land; the house was demolished in 1937 and the estate used for housing. The Clock Tower was sold too in 1938 and is now a landmark in the centre of Warsash. He moved to Winkworth Hall, Hascombe, Godalming, where he died on 21 July 1961.

The Royal Aero Club dinner for pioneers held on 26 November 1957: from left to right, Lord Brabazon, Fred Handley Page, Geoffrey de Havilland, the Duke of Edinburgh, Claude Grahame-White and TOM Sopwith.

ACKNOWLEDGEMENTS

I began this work as research to enable the accurate cataloguing of photographs in the museum's collection. When they saw the work I was doing my colleagues encouraged me to write a new history of Hendon and it is to them I offer my thanks for their encouragement and patience while my time has been occupied by the task. My thanks are also due to previous authors who have written about the aerodrome from whom I could draw inspiration and guidance.

The many men and women who have worked or visited Hendon have also been of great assistance through their writing and photographs, not least Claude Grahame-White, T Kemp Walton, Richard Gates and Marcus Manton, material which their families kindly made available to the museum. Material has also been obtained from the Royal Aero Club whose collection is kept at Hendon. Hugh Petrie of Barnet Archives with his encyclopaedic knowledge of Hendon has also been of great assistance.

Previous authors lacked access to the wealth of material being placed on-line. The staff and partners of the American Library of Congress, The National Archives at Kew, the New York Times and Flight International among others have helped make research far easier than in the past.

Andrew Renwick

INDEX

A Harper, Sons & Bean Ltd 64-65
Abbey Road, London 86
Aden, British evacuation from 97, 99
Aerial League of the British Empire 16
Aerial Propulsion Syndicate Ltd 21, 23
Aero Club of Great Britain and Ireland
 See Royal Aero Club

Aerodromes
 Acton 46
 Bassingbourn 91
 Beaulieu, Hampshire 13
 Benson 86
 Biggin Hill 82
 Blackbushe 100
 Bovingdon 89, 91, 99
 Broadwell 87
 Catterick 84
 Chelmsford 42
 Chingford 42-44, 48, 57
 Cosford 85
 Croydon 68, 71, 86
 Doncaster 88
 Eastchurch 24
 Eglinton 92
 Gatwick 99
 Hatfield 86
 Heathrow 93, 96
 Henlow 101
 Heston 86-87, 108
 Hornchurch 96
 Kemble 85, 88
 Kenley 53
 Lanark 29
 Larkhill 21
 Leavesden 91
 Leconfield 86
 Long Kesh 90, 92
 Lympne 51, 71, 75
 Lyneham 87
 North Weald 82, 96
 Northolt 74-75, 82, 84, 96, 99-100
 Park Royal 14-16, 19
 Pau 11-14, 18
 Port Lyautey 92-94
 Prestwick 87
 Redhill 87
 Shoreham 24
 St Mawgan 87
 Stag Lane 46, 51
 Stapleford Tawney 86
 Thorney Island 84
 Waddon See Croydon
 Warton 87
 West Raynham 87
 White Waltham 81

Aerofilms Ltd 65-66
Aero-Marine Engines Ltd 108
Aeronautical Inspection Department 48-52
Aeronautical Syndicate Ltd 21, 23-24, 27, 29
African Aviation Syndicate Ltd 26
Agricultural Halls, Islington, London 53
AGS Bond, West Hendon 52
Air Historical Branch, Ministry of Defence 96, 101
Air Mail, first UK service Hendon-Windsor 25
Air Ministry Mechanical Transport Section 82
Aircraft Co, The See Aircraft Manufacturing Co Ltd
Aircraft Disposal Co Ltd 53
Aircraft Manufacturing Co Ltd 27, 29, 34, 59, 65-66
Airships Ltd 34-35
Alfred Dugdale Ltd 71-72, 76

Angus Sanderson Ltd 70, 72
Apprentices 33-34
Ardern, Lawrence 20
Arkwright, Harold Arthur 20, 25
Armistice 11 November 1918, effects of 61
Arnell, R S 46
Austin Motor Co Ltd 59
Austin, Herbert 9
Automobile Club of Great Britain See Royal
 Automobile Club
Avenue Road, Hendon 18
Aviation Investment & Research Ltd 16
Bailhache Committee of Inquiry, 1916 49
Baker, Captain Valentine Henry 108
Baldwin, Stanley, 1st Earl Baldwin of Bewdley 60
Balfour, Arthur James, 1st Earl of Balfour 9, 36
Balloon Society of Great Britain 7
Ballooning 6-7, 11
Barber, Horatio 21, 23, 27
Barham, Sir George 26
Barnet Council 100
Barrie, Sir James Matthew 17
Barrs, Alfred Edward 54
Battle of Britain 84-85
Bauman, Eric Bentley 38
Baumann, Édouard 29, 41
Bean See A Harper, Sons & Bean Ltd
Beatty Aviation Co Ltd 35, 45-47, 54
Beatty School See Beatty Aviation Co Ltd
Beatty, George William 35
Beaufort Park Estate, Colindale 104-105
Beaumont, André See Conneau, Lieutenant de Vaisseau
 Jean
Bedford & District Model Aeroplane Club 108
Bedford Grammar School 9
Belgium, RNAS attacks on 39-40
Bey, Henry 72
Biard, Henry Charles 38, 47
Birchenough, William 36
Bird Estate, Colindale 102
Blackburn School of Flying 29
Blackburn, Harold 29
Blair Atholl Aeroplane Syndicate Ltd 32
Bleriot Aeronautics 20-22
Bleriot School of Flying See Bleriot Aeronautics
Blériot, Louis 11, 25, 34
Bliss, Miss Ellen Pauline Matthew 17, 21
Book Production & Distribution Centre 88-89
Boudot, E 54, 59, 63
Boyle, Marshal of the Royal Air Force Sir Dermot 103
Brackenridge, Henry Philip Gerald 45, 47, 68
Breguet Aeroplanes Ltd 29, 32
Brewer, Cecil and Maurice 26
Brewer, Robert Wellesley Anthony 13-15
Brinckman, Colonel Sir Theodore 26
British & Colonial Aeroplane Co Ltd 18, 22
British Aerial Transport Co Ltd 61, 66
British Army 83, 90, 93
British Caudron Co Ltd 34, 41
British Deperdussin Aeroplane Co Ltd 29, 33, 39
British Deperdussin Aeroplane Syndicate Ltd 29
British European Airways 96
British Gliding Association 108
British Library Newspapers, Colindale 105
British Motor Boat Club 10, 68, 106
British Trans-Atlantic Flight Fund 36
Brock, Henry le Marchant 28
Brodribb, Squadron Commander Francis George 50
Brooklands, motor racing at 10
Broughton, Air Marshal Sir Charles 101
Brown, Michael 100

Brytcast Stainless Metals Ltd 107
Buckingham Palace, London 84-85, 93
Bursledon Towers, Southampton 8
Busteed, Flight Commander Henry Richard 50
Butler, Frank Hedges 20
C Grahame-White & Co Ltd 11, 21, 25-26
Cambrian Coaching & Goods Transport Co Ltd 71-72
Cambridge School of Flying & Aerodrome Co Ltd 45
Cammell, Lieutenant Reginald Archibald 24
Canada, presentation of aircraft to 53
Canberra House 74, 78, 91
Capazza, Louis Henri 7
Carr, Reginald Hugh 33, 63
Casualties, air evacuation of 86-87
Catapults, used for experimental launching of aircraft
 50
Cedars Close, Hendon 87-88, 91
Chamberlayne, Air Commodore Paul Richard
 Tankerville James Michael Isidore Camille See
 Tankerville-Chamberlayne, Captain Paul Richard
Chanter School of Flying 24
Chanter, M 24, 29
Chase, Miss Pauline See Bliss, Ellen Pauline Matthew
Cheltenham & West of England Aviation Co Ltd 45
Chereau, Norbert 20
Chinnock, Frederick 8
Chinnock, Miss Ada Beatrice 8, 14, 73
Chinnock, Miss Florence 8
Christie, Dr Malcolm Grahame 29
Church Farm, Hendon 26, 56
Churchill, Sir Winston 38, 57
Cinema 97
Clapham Properties Ltd 109
Clitterhouse Farm, Cricklewood 46
Clutterbuck, E C 21
Code Books, production of 88-89
Cohen, George M 56
Cohen, Miss Georgia Ethelia (Georgette) 56, 66-67
Colindale Area Action Plan 105
Colindale Hospital 89, 105
Colindale Police Station 102
Colindale Underground Station 85, 105
Computers, use of by the RAF 97, 100, 102
Conneau, Lieutenant de Vaisseau Jean 24
Conscription, effects of 41, 45, 58
Cook, Miss Edith Maud 13-14
Courtney, Frank Thomas 34
Cowes, racing at 10
Coxwell, Henry Tracey 6
Cricklewood, London 41, 46
Crondall School, Farnham, Surrey 9
Crosbie, Wing Commander Dudley Stuart Kays 75
Crystal Palace, London 16, 25
Curtis, Miss Lettice 92
Curtiss, Glenn 18, 30
Daily Mail London-Manchester Race, 1910 7, 14-16,
 19
De Havilland, Sir Geoffrey 34, 109
Defence of the Realm Act, land requisitioned under 50
Delco Lighting 72
Dentice de Frasso, Count Carlo 56
Deperdussin School of Flying See British Deperdussin
Deperdussin, Armand 33
Dowding, Air Chief Marshal Hugh Caswall
 Tremenheere 79
Drexel, John Armstrong 13, 30
Driver, Evelyn Frederick 25-26
Drummond, Captain Alexander Victor 22
Dunne, Lieutenant John William 32
Dyott, George Miller 32
Eastbourne Flying Club 108

Eastchurch Naval Flying School RNAS 39
Edgcumbe, Major Kenelm William Edward, 6th Earl of Mount Edgcumbe 19-20
Egypt, Anglo-French invasion of, 1956 95
England, flooding of the east coast of, 1953 95
Esnault -Pelterie, Robert 73
Everett Edgcumbe & Co Ltd 19, 26
Everett Edgcumbe Monoplane 19-21
Everett, Edgar Isaac 19, 26
Ewen School of Flying 29, 34
Ewen, William Hugh 29
Fairey, Charles Richard 19, 31
Farman, Henry 15
Faygate, Surrey 109
Featherstone Farm, Mill Hill 26, 56, 74, 76
Fielden, Wing Commander Edward Hedley 79
Finchley Football Club 74
Fitzsimmons, John Bernard 55
Fletcher, Anthony A 54
Forrest, Miss Elsie See Tikrani, Maharanee of
France, Fall of 83
Franco British Electrical Co Ltd 72, 74, 76
Franco-British Exhibition 1908 72
Fraser, Alexander 71
Fraser's Flying School 71
Fulham Broadway, London 11, 13
Fulton, Captain John Duncan Bertie 13
Gage, Captain James Seton Drummond 109
Gar Wood Motor Boats Ltd 106-107, 109
Garnier, R 29
Gates, Richard Thomas 25, 38-40
Gathergood, Captain Gerald William 65
Gaunt, W C 68, 72
General Motors Ltd 71-72
Geraldo & His Orchestra 89
Glaisher, James 6
Gliding 96, 108
Gooch, Sir Daniel 29
Gordon Bennett Cup for Aeroplanes, 1910 18, 24
Gordon Bennett Cup for Aeroplanes, 1911 24
Gordon Bennett Cup for Balloons, 1906 11
Gordon Bennett Cup for Cars, 1902 9
Grahame Park Estate, Colindale 100, 105
Grahame-White Aviation Co Ltd 25-27, 56-69, 72-73
Grahame-White Co Ltd See Grahame-White Aviation Co Ltd
Grahame-White School of Flying 13-14, 16, 18, 21, 25, 33, 38, 42-43, 45-47, 57, 65
Grahame-White, Bleriot & Maxim Ltd 25
Grahame-White, John Reginald 8-9, 11
Grahame-White, Miss Beatrice Eleanor Genevieve 8-9, 11, 15, 109
Grahame-White, Montague Reginald 8-9, 15, 107, 109
Grapperon, André 18
Greswell, Clement Hugh 21, 33
Greyhound Inn, Hendon 6, 34
Grimmer, Robert Paul See Mann & Grimmer
Gunter, Robert W 36
Halifax, West Yorkshire 16
Hall School of Flying 32, 45-46, 54
Hall, John Lawrence 32, 41, 45
Hallam, Lieutenant Theodore Douglas 50
Hamel, Gustav 25, 30, 34, 40
Handasyde, George Harris 7
Handley Page School of Flying 35
Handley Page, Sir Fred 27, 109
Harmsworth, Alfred Charles William, 1st Viscount Northcliffe 106
Harper, Harry 6
Harrison, Stanley Price Knowles 71
Heal, John 26
Henderson, John S C 50
Hendon Hall Hotel 87-89
Hendon Urban District Council 19, 25, 71
Hickman Sea Sled 67, 69
Hickman, William Albert 67
Hill, Cecil McKenzie 45-46
Hillfoot House, Hendon 74, 78, 91
Hinds-Howell, Captain Geoffrey Llewellyn 21
Hinkler, Herbert John Louis (Bert) 76
HM King Edward VIII 77-79
HM King George V 25, 50, 60, 79
HM King George VI 79, 90, 95
HM Queen Elizabeth II 90, 101, 103
HM Queen Elizabeth, The Queen Mother 90, 102-103
Hoare, Sir Samuel 74
Holmes, Sergeant Raymond T 84
Honeyman, David Livingstone 72
Hornby, Lieutenant Commander Christopher 40, 48
Hotel Trust Ltd 109
HRH Prince Charles, Prince of Wales 102
HRH Prince Edward, Prince of Wales See King Edward VIII
HRH Prince George, Duke of Kent 86
HRH Prince Philip, Duke of Edinburgh 102-104, 109
Hubert, Charles L A 30
Hucks, Bentfield Charles 69
Hucks, Frank 30
Hunter, W D (Doug) 33
Hutchins, Harry Easdown 68, 74
Hythe, motor boat racing at 106
Ideal Home Exhibition 1920 63
Industrial Guarantee Corporation Ltd 70
Innes-Ker, Major Lord Robert 50
Irvine, Miss Lily 22
Isaac, Bernard 44, 68
Johns, Captain William Earl 76
Johnson, Claude 9
Johnson, William Lyulph 46
Kavanagh, Miss Spencer See Cook, Edith Maud
King, E Warr 63
Leary, George Jnr 67
Lee, Miss Phoebe 108
Levey, Miss Ethel Grace 56, 66-67, 107-108
Levitt, Miss Dorothy 106
Liquid Fuel Engineering Co Ltd 10
Liverpool Motor House Ltd 26
Livingston, Guy 26
Lloyd's of London 71
London & Provincial Aviation Co Ltd 35, 45-46, 54
London & Provincial School of Flying See London & Provincial Aviation Co Ltd
London Aerodrome Club Ltd 61
London Aerodrome Ltd 20, 25-26
London Aviation Ground Ltd 46
London Central Garage Co Ltd 109
London Country Club Ltd 61, 65-66, 73
London Flying Club Ltd See London Country Club Ltd
London Flying Society 72
London Underground 68, 76
London, air defence of 42-44, 50, 84-86, 97
London, simulated attacks on 97
London-Brighton Emancipation Run, 1896 9
MacFie, Robert 19
Malta, re-enforcement of 86
Mander, Lionel Henry (Miles) 14
Mann & Grimmer 54-55
Mann, Reginald See Mann & Grimmer
Mansfield, George Harold 26
Manton, Marcus Dyce 33, 38, 42
Maple, Sir John Blundell 29, 68
Marble Arch Line, US Air Transport Command service 87
Marconi, Guillermo 106
Markham, Charles 107
Martin, Helmuth Paul 7
Martin, James Vernon 24, 36
Martlesham Heath 83
Matthews, Herbert William 25, 59, 65
Maxim, Sir Hiram 16, 25
McArdle, William Edward 13
Meering, Frank Clement 47
Merriam, Frederick Walter 42-44
Meteorological Office 97
Metropolitan Police Training College 78, 90, 105
Middlesex University 102, 104
Miles, Wing Commander Charles Cleaver 78

Military Units, American
Air Transport Command 87, 89
Fleet Aircraft Service Squadron FASRON (Special) 200 94-95
Fleet Aircraft Service Squadron FASRON 76 94
Fleet Aircraft Service Squadron FASRON 77 94
VR-24 93-94
VR-25 94
VRU-4 92-93
VIII Air Force 87
XII Air Force 87
27th Air Transport Group 87
32nd Anti-Aircraft Artillery Brigade 95
2008th Transport Group (Provisional) 87
86th Air Transport Squadron 87-88
87th Air Transport Squadron 87

Military Units, British
A (Temporary) Maintenance Depot 81
A Equipment Depot 78, 81
Advanced Air Striking Force 82-83
Air Despatch & Reception Unit 90
Air Despatch Letter Service See 1697 (Air Delivery Letter Service) Flight
Air Trooping & Freight Control Section 94
Anti-Aircraft Flight 86
Army School of Physical Training 90
Autogiro Unit, The 83
Auxiliary Air Force See Royal Auxiliary Air Force
Blenheim Conversion Flight 81-83
British Expeditionary Force 82-83
Civilian Aerial Erection Flight 96, 103
Coastal Command Communications Squadron 91
Dakota Conversion Section 88
Exhibition Design & Display Unit 95-96, 101, 103
Fighter Command Communications Flight 91-92
Hendon Aircraft Acceptance Park See 2 (Hendon) Aircraft Acceptance Park
Hendon Aircraft Production Depot 52
Hendon Communications Flight See 1 (Communications) Squadron
Home Command See Reserve Command
Home Communications Flight 78
Joint Services Air Trooping Centre 94-96, 99-100
King's Flight 79, 81-83
Medical Flight See 29 Training Squadron
Metropolitan Communications Squadron 89-91, 94-96
Naval Flying School 42-43, 48
Passenger Handling Section See Air Trooping & Freight Control Section
Radio Fitting Unit 85
Radio Installation (Mobile) Section 85
Refuelling & Re-arming Party 86
Reserve Command 81, 93
Royal Air Force Antarctic Flight 93
Royal Air Force Cinema Corporation 96-97
Royal Air Force Despatch Rider Letter Service 89
Royal Air Force Regiment 86
Royal Air Force Volunteer Reserve 79, 93
Royal Auxiliary Air Force 74-76, 79, 81-82, 93
Royal Flying Corps School of Instruction 45-47
Royal Naval Volunteer Reserve 109
Special Fighter Cadre Unit 82
Supply Control & Accounting Flight 104
Supply Control Centre 97-98, 100-101

Military Units, British continued
Transport Command Communications Flight 89
Women's Auxiliary Air Force 87-89, 91
Women's Royal Air Force 94
1 (Communications) Squadron 50-51, 53
1 Aircraft (Salvage) Depot See Hendon Aircraft Production Depot
1 Aircraft Delivery Flight 86
1 Camouflage Unit 85-86
1 Casualty Air Evacuation Unit 87
1 Mechanical Transport Storage Depot 78
2 (Hendon) Aircraft Acceptance Park 46-47, 50-53
2 Air Formation Signals Unit 83
2 Casualty Air Evacuation Unit 87
3 Casualty Air Evacuation Unit 87
3 W/T Transportable Unit 83
4 Motor Transport Unit 86
13 Air Formation Signals Unit 83
17 Reserve Squadron RFC 44
19 Reserve Squadron RFC 44
24 squadron 78, 81-82, 85-89, 91
26 (Training) Group 79
29 Training Squadron 51
31 Squadron 92, 94
47 Group 90-91
50 Group 81
52 Wing 83
59 Squadron 84
62 Fighter Servicing Wing 82-83
65 Group 93
67 Fighter Servicing Wing 83
81 Squadron 83
93 (Forward) Staging Post 87
107 Wing 89
116 Squadron 86
116 Wing 88-90
120 Squadron Air training Corps 100
142 Gliding School 96
151 Wing 86
248 Squadron 84
257 Squadron 84
500 Squadron 74
504 Squadron 84
510 Squadron 87, 89
512 Squadron 87
575 Squadron 87
600 Squadron 74-76, 79, 81-82
601 Squadron 74-75, 77, 79, 81-82, 93
604 Squadron 76, 79, 81-82, 93
617 Gliding School 96-97, 99
661 Auxiliary Air Observation Squadron 93-94
705 Signals Unit 97
1316 Flight 89
1416 Flight 86
1697 (Air Delivery Letter Service) Flight 82, 89
1958 Auxiliary Air Observation Flight 93
2722 Squadron RAF Regiment 86
4024 Servicing Echelon 91

Mill Hill, London 6
Ministry of Munitions 49, 53, 55, 60
Moore-Brabazon, John Theodore Cuthbert, 1st Baron Brabazon of Tara 16, 108-109
Mountbatten, Louis, 1st Earl Mountbatten of Burma 107
Munich Crisis 81
Nash Collection 96, 101
Nestler Ltd 54-55
New Hendon Village, Colindale 105
New Zealand, flying training in 46
Newell, William 33
Nieuport (England) Ltd 32
Noel, Louis 33, 38
Normandy, preparations for invasion of 88
North, John Dudley 27, 30, 59
Northern Aero Syndicate 16
Norway, RAF units in 83
Nungesser, Lieutenant Charles Eugène Jules Marie 65
O'Gorman, Mervyn Joseph Pius 30
Operation Dunlop 86
Operation King Canute 95
Operation Musketeer See Egypt, Anglo-French invasion of, 1956
Operation Sealion 84, 86
Orange Hill House, Edgware 29
Osipenko, Mavreky 47
Parachuting 13, 33
Parliamentary Aerial Defence Committee demonstration, 1911 22-23
Pashley, Cecil Lawrence 47, 72
Paterson. Charles Compton 26
Paulhan, Louis 7, 14-16, 18-19, 30
Payne, Frederick Harrold 56, 59
Peel Centre See Metropolitan Police Training College
Penny, Flight Commander Rupert E 50
Pickles, Sydney 28-29, 45
Piggots Manor, Letchmore Heath 56
Porte, Lieutenant Commander John Cyril 28-29, 39-40, 42
Portman Trust Ltd 109
Prier, Pierre 20, 22
Prosser, Edwin T 41
Pulse Development, Colindale See Colindale Hospital
Pupin, Emile 21
Ramsay-Fairfax, Lieutenant George A 25-26
Reader, Ralph 90
Reyrol Motor Car Co Ltd 10
Richard, Major Leslie Fitzroy 45, 47
Ridley-Prentice, W 23
Risk, Wing Commander Charles Erskine 50
Ritchie, Thomas E 59
Road Research Laboratories 97
Ross, Hamilton 24
Rossmore Court, London 107
Rouse, William Hamilton 67
Royal Aero Club 11, 13, 24, 29, 44-45, 65, 69, 72, 74, 94, 106, 108-109
Royal Aeronautical Society 96
Royal Air Force, formation of, 1 April 1918 51
Royal Automobile Club 9
Royal Motor Yacht Club 68
Royalty 25, 50, 60, 77-79, 83, 86, 90, 95, 101-104
Ruffy, Arnell & Baumann Aviation Co Ltd 41, 45-46, 54, 61
Ruffy, Félix 41, 45-46
Ruffy-Baumann School See Ruffy, Arnell & Baumann Aviation Co Ltd
Ruislip, Middlesex 94
Salmet, Henri 22, 36
Santoni, D Lawrence 33
Santos-Dumont, Alberto 30
Sassoon, Sir Philip 75
Satterthwaite, Livingston 92
Savage Sky-Writing Co 68-69, 72, 74, 76, 96
Savage, Major John Clifford (Jack) 68-69
Shackel, Miss Ellen 8
Shaw, George Bernard 44
Shipley Wool Combing Co 9
Silver Thimble Fund 86
Simpson, Wing Commander William George 102
Smiles, Michael Geoffrey 35
Smiles, Walter Dorling 35
Smith, Richard Tilden 68
Society of British Aircraft Constructors 56, 80
Soldiers, Sailors, Airmen & Families Association 99
Sopwith, Sir Thomas Octave Murdoch 29, 109
South Africa, early aviation in 26
Souther, J William 66-67
Southport 21, 24, 26
Soviet Union, RAF assistance to 86
Spencer, Percival 6-7
Sports, aircraft workers' 59
St Mary's Church, Hendon 34, 39-40
Stanbridge 102
Standard Telephones & Cables Co Ltd 72-74, 76, 78
Stanmore 96
Stanstead Park, Surrey 10
Stewart, Gordon 21
Sueter, Rear Admiral Sir Murray Fraser 38-39
Surplus aircraft and vehicles, disposal of, 1919 53
Sutherland, Miss Florrie 'Birdie' See Watkins, Annie Louise
Taft, William Howard 17-18
Tankerville-Chamberlayne, Captain Paul Richard 65
Tanner, John 101, 103
Taylor, Miss Dorothy Cadwell 29-31, 56
Temple School of Flying 29, 32
Temple, Colonel Richard Durand 70-71
Temple, George Lee 29
Tetard, Maurice 18
Theatrical actresses and productions 9-10, 17, 21, 36, 56, 66-67, 89, 107-109
Thomas, George Holt 15-16, 27, 59
Thousand Mile Reliability Trial, 1900 9
Tikrani, Maharanee of 36
Tithe Farm, Hendon 26
Trams, service extended to Edgware 12
Turner, Lewis William Francis 34
Tylor Engineering Co Ltd 69-70, 72
United States Air Force 99
United States Army 95
United States Navy 92-95
United States of America Air Attaché Flight 92
United States of America, entry into World War Two 86-87
Uphill Road, Mill Hill 88-89
Vaulx, Henri de la 11
Vedrines, Jules 24
Verrier, Pierre 30, 36
Victory in Europe 1945, effects of 90-91
V-weapons, attacks on England by 89-90
W H Ewen Aviation Co Ltd See Ewen School of Flying
Wallace, Grahame 9
Walton, T Kemp 59
Warburton, Barclay Harding Jnr 65
Warren, William Thomas 35-36
Warsash 109
Watkins, Miss Annie Louise 9, 109
Watling Estate, Colindale 76
Webb, Air Mechanic Reggie 42
Wells Aviation Co 45
Welsh Harp, London 6-7, 12-13, 19, 26
Wembley, London 81
West German Air Force 99
Wilder, George 10
Willesden, London 29, 61
Willey, Francis, 1st Baron Barnby 8-9, 68
William Cole & Sons (1923) Ltd 71
Williams, Leonard 63, 68
Williamson, Leonard 24
Willows, Ernest Thompson 34-35
Wills, Francis Lewis 65
Winchester, Clarence (Ornis) 65, 72
Windham, Captain Walter George 24
Winkworth Hall, Godalming, Surrey 109
Winter, John Scott Bradbury 47
Wood, Garfield Arthur 106-107
World War One, start of 36, 38
World War Two, start of 82
Wormwood Scrubbs, London 14-16
Wright Co 18, 24
Wright, Wilbur and Orville 7, 18, 30
Wurtemburg, Crown Prince of See Gunter, Robert W
Wycombe Aircraft Constructors Ltd 59
Yorkshire Motor Vehicle Co 10